SOLID WASTE: NEW RESEARCH

SOLID WASTE: NEW RESEARCH

ADRIAN S. MARTIN
AND
JAVIER D. SUAREZ
EDITORS

Nova Science Publishers, Inc.
New York

For permission to use material from this book please contact us:
Telephone 631-231-7269; Fax 631-231-8175
Web Site: http://www.novapublishers.com

LIBRARY OF CONGRESS CATALOGING-IN-PUBLICATION DATA

Martin, Adrian.
Solid waste : new research / Adrian S. Martin and Javier D. Suarez, editors.
p. cm.
Includes bibliographical references and index.
ISBN 978-1-60456-809-7 (hardcover)
1. Refuse and refuse disposal--Research. I. Suarez, Javier. II. Title.
TD793.3.M37 2008
628'.744--dc22

2008023916

Published by Nova Science Publishers, Inc. ✢ New York

Contents

PREFACE

Solid waste is any garbage or refuse from industrial, commercial, agricultural or community activities. This book provides leading edge research on this broad field of study from around the globe.

Chapter 1 - Waste to energy has a clear dual role in the context of sustainable waste management: on one hand, it has to reduce the volume and hazardousness of wastes that can receive no further treatment prior to disposal; on the other, it has to make the most out of the energy content of the treated wastes, in order to improve the economics and environmental performance of the whole process. This paper presents an overview of the state-of-the-art in this field, highlighting some related trends, needs, barriers, opportunities, drivers, pros and contras. The surveyed technologies fall under the following categorization:

Conventional incineration of mixed wastes and residues with energy recovery and flue-gas treatment.

Pyrolysis and gasification of mixed wastes, pre-treated fractions and residues.

Co-firing of waste derived fuels in industrial furnaces.

Anaerobic digestion of organic substrates and mixed wastes/residues in digesters.

Thermal inertization / detoxification of hazardous waste fractions.

Pretreatment of wastes for subsequent thermal treatment and energy recovery; waste derived fuels (refuse derived fuel, tire derived fuel, packaging derived fuel, solid recovered fuel).

Chapter 2 - Almost all the metallurgical processes are associated with the generation of wastes/residues which may be hazardous or non-hazardous in nature depending upon the criteria specified by institutions like US EPA etc. The wastes containing heavy and toxic metals such as arsenic, cadmium, chromium, nickel, lead, copper, mercury, zinc etc. are present beyond permissible limits deemed to be disposed off. Due to the implementation of stricter environmental laws and economic reasons all the metallurgical industries are now forced to go for eco-friendly technologies to produce metal and other related products world over. However, generation of wastes is the integral part of metallurgical industries which can not be ruled out, therefore if the wastes/residues are hazardous in nature they generally have to be treated or disposed off in safe and designated dumping sites. If these wastes/residues are non-hazardous in nature then they may be suitably used as secondary raw material for the recovery of metals which are in growing demand all over the world. Zinc is in growing demand all over the world. In India a major amount of zinc is imported and therefore processing of zinc secondaries will supplement in satisfying the gap between demand and supply to some extent. Similarly processing of lead secondaries is important because of their relative high metal content, besides low energy and cost involved in recovering the metal. This chapter highlights the production capacity, type and quantity of solid wastes generated, their chemical composition and treatment/disposal options for the Indian lead and zinc industries. Zinc tailing, slag, leach residue, jarosite residue, β-cake, etc. from zinc industries and BF slag, flue dust, ISF slag etc. from lead industries are the major solid waste generated from various process and needs attention. Although all the metal producing industries in organised sector are now taking care of the environment and waste management related problems, but pollution from unorganized lead units are the major cause of concern. Permissible limits of toxic constituents in zinc based secondaries and threshold zinc concentration for both indigenous and imported raw materials were worked out at National Metallurgical Laboratory (NML). An overview of the current practices and recent trends in the secondary processing of zinc and lead and the attempts made to recycle/recover metal values and production of value added products, are discussed in the text. Various processes, particularly hydrometallurgical ones, already developed or in the developmental stages, are discussed. Attempts made by various laboratories and industries towards the development of eco-friendly processes for the recovery of zinc and lead from secondary raw material are also described.

Chapter 3 - This work presents an analysis of municipal solid waste management, from an economic perspective. The authors discuss the market failures caused by generation and management of waste and the main policy instruments to deal with those failures. Specifically, they focus on incentive policies and we list the main advantages and disadvantages of each instrument. The authors also present a brief summary about the recent history concerning both the principles and practices for waste management in the international context.

Chapter 4 - This report, which replaces a 2004 report on the same subject (CRS Report RL32570), provides updated information on interstate shipment of municipal solid waste (MSW). Since the late 1980s, Congress has considered, but not enacted, numerous bills that would allow states to impose restrictions on interstate waste shipments, a step the Constitution prohibits in the absence of congressional authorization. Over this period, there has been a continuing interest in knowing how much waste is being shipped across state lines for disposal, and what states might be affected by proposed legislation. This report provides data useful in addressing these questions. It generally presents data as of 2005.

Total interstate waste shipments continue to rise due to the closure of older local landfills and the consolidation of the waste management industry. More than 42 million tons of municipal solid waste crossed state lines for disposal in 2005, an increase of 8% over 2003. Waste imports have grown significantly since CRS began tracking them in the early 1990s, and now represent 25.3% of the municipal solid waste disposed at landfills and waste combustion facilities. In the last 10 years, reported imports have increased 147%.

Pennsylvania remains the largest waste importer. The state received more than 7.9 million tons of MSW and 1.7 million tons of other non-hazardous waste from out of state in 2005. Most of this waste came from New Jersey and New York. Pennsylvania's waste imports represented 19% of the national total. Virginia and Michigan, the second and third largest importers, received 5.7 million tons and 5.4 million tons from out of state respectively in 2005, each of them about 30% less than the amount received by Pennsylvania.

With the exception of Pennsylvania, each of the 15 largest importers showed an increase in waste imports, compared to our last survey, which provided data as of 2003. Indiana, Michigan, and Wisconsin showed particularly large increases, with Ohio, New York, Oregon, and Georgia also increasing substantially. In each of these states, waste imports increased by 300,000 tons or more, in some cases substantially more. In all, 30 states had increased imports in the current report, and 11 states reported imports that exceeded 1 million tons.

While waste imports increased overall, Pennsylvania, the leading importer, reported a sharp decline in imports. Pennsylvania's imports fell for the fourth year in a row: about 2.7 million fewer tons of out-of-state MSW were received at Pennsylvania landfills in 2005 than in 2001. Factors causing this decline included the imposition of an additional $4.00 per ton state fee on waste disposal and the absence of rail service at Pennsylvania landfills.

New York remains the largest *exporter* of waste, with New Jersey in second place. Nine other states (Illinois, Missouri, Maryland, Massachusetts, Washington, Minnesota, North Carolina, Indiana, and Florida), the District of Columbia, and the Canadian province of Ontario also exported more than 1 million tons each.

Chapter 5 - Usually, the Municipal Solid Waste Incineration (MSWI) fly ashes are solidified/stabilized (S/S) in cement and disposed in appropriate landfills, producing huge monoliths. In order to understand the chemical reactions, which can take place inside such monoliths, an analytical approach at the micron and even at the infra-micron scale is required. Considering, this approach, which practically combines wet chemical analyses, leaching tests, XRD and SEM-EDS characterization, samples of naturally aged and fresh S/S MSWI fly ashes, were examined to describe and to predict the long-term behavior of lead. The compliance leaching tests indicate that considerable amounts of lead are leached for several samples. In addition to portlandite and calcite, the main major minerals identified are hydrocalumite, ettringite and ordered silicate-hydrates, the three later being free of lead in detectable concentrations. The lead speciation shows a rather large variety of minor Pb-bearing compounds, including soluble minerals and amorphous phases. As observed by SEM performed on polished thin sections, the neo-formed Pb-hydroxychlorides are partly related to a concentric sequence involving the aluminum hydrolysis, actually aluminum (internal) / aluminum hydroxide (intermediate) / hydrocalumite (external). Pb-minerals, typically cotunnite, blixite, laurionite, penfieldite or Pb-hydroxides are visible inside the hydrocalumite zone. Combined with the high Cl^- activity derived from the MSWI fly ashes, the aluminum hydrolysis occurring in cementing conditions partly governs the status of lead by promoting the precipitation of unstable neo-formed Pb-minerals.

Chapter 6 - Growing production of municipal solid waste in the recent years caused by increasing global levels of economic activity needs more sophisticated management tools for its collection. Despite some local efforts to reduce the waste, disposal in landfills is still the most predominate way. Considering to restrictive environmental regulations, more efficient and sustainable systems of municipal solid waste management are carried out to protect our living environment. They deal with the waste sorting, which is mainly focused on categories represented by paper, mixed plastics, mixed glass, mixed waste, hazardous waste and metals. Thus, specific transport vehicles and optimized routes are needed to manage the collection of all the waste categories from local sites to landfills, incineration plants, collecting yards and recovery sites. This contribution presents the waste collection methods based on spatial analysis and operations research. The spatial analysis is focused on development of the road network datasets for waste collection, finding the best routes, defining the service area and creating the origin-destination cost matrix in geographic information systems (GISs). The identified routes and local service areas together with the cost per unit and the total amount of the transported waste form data inputs into operations research. The transportation problem in linear programming is used to optimize the total cost for transportation of each waste category in the framework of municipal solid waste collection. As an example, the urban areas of Prague are included into a case study to illustrate spatial modeling in the GIS environment and optimization focused on collection networks. The individual collection sites are grouped into the urban areas. Waste production in each urban area is estimated in dependence on available data from reports about the living environment in Prague. The locations of target collection points represented by recycling places, incineration plants and landfills are based on larger urban districts. The developed GIS project can manage various scenarios, which indicate different total costs. In spite of its actual lower total cost, the landfill will become more difficult to implement due to the increasing cost caused by limited land availability and environmental community opposition. On the other hand, the recovery/recycling will become more important because of its minimal negative interaction with the living environment. In order to minimize the costs and degradation of our living environment, spatial modeling extended by methods focused on linear programming can assist to find suitable waste management scenarios for collection of municipal solid waste in urban areas.

Chapter 7 - The purpose of this work is to evaluate the influence of the principal intervening factors in anaerobic treatment process of organic solid waste. The results obtained from the research work on biological treatment of organic solid waste in anaerobic reactors which has been realized since 1997 by the Environmental Sanitation Group of Universidade Estadual da Paraíba, northeast of Brazil will be presented. During these years various operational parameters such as: size and configuration of the reactor, moisture amount, type and amount of inoculum, and others have been studied. Treatments with high and low concentrations of solids were realized and the results obtained showed that the removal efficiency of the applied mass to the reactor and the moisture amount are directly proportional. During the anaerobic treatment with high concentration of solids, one problem that can be verified was the high concentration of volatile fatty acids and chemical oxygen demand in the liquid leachate, which tends to lower the pH down to the levels between 2.0 to 4.0, thus inhibiting the methanogenic microorganisms. During the process of anaerobic treatment with high concentration of solids the biogas produced were almost totally composed of CO_2. When the process attains the dynamic equilibrium the rate of CH_4 production attains the maximum, then it decreases continuously through the end of the reactor operation. In the case of anaerobic treatment with low concentration of solids, the biogas production is practically constant after attaining the equilibrium in the reactor. Gas chromatographic analysis of the biogas showed that the amount of methane went as high as 70%, depending upon the applied operational parameters. Bovine rumen, UASB sludge and leachate were used as inoculum, and it was verified that in general the biostabilization time of the waste was inversely proportional to the percentage of inoculum used, and independent of the inoculum type. Scanning electron microscopic and fluorescence optical microscopic analyses of the samples from the reactors showed predominance of *Methanosarcina sp.* kind of microorganisms, which explains the high amount of methane in the biogas. In general, it can be concluded that the various intervening factors existing in the organic solid waste treatment process in anaerobic reactors have made difficult the application of this type of process at real scale. Thus, one should be very careful while projecting this type of treatment system.

Chapter 8 - This article investigated the properties of municipal solid waste incinerator (MSWI) ashes and the effects of their addition on the municipal solid waste (MSW) anaerobic digestion as co-disposed or co-digested with MSW in landfill or digester. Five anaerobic bioreactors with the size of 1.2 m height and 0.2 m diameter were employed to conduct the experiment. Four layers were arranged each with 6.5 liter of MSW and anaerobic seeded sludge mixture covered by 2.5 liter of MSW and anaerobic seeded sludge mixture blended with the designed ashes added ratios as well as the control bioreactors without ashes addition. The synthetic MSW used in this experiment was typical of organic fraction of MSW and was comprised of newspaper, food waste, office paper and hay etc. MSWI ashes were obtained from a mass burning incinerator in central Taiwan. Also, the seeded anaerobic sludge was taken from a municipal wastewater treatment in central Taiwan. As the experiment was proceeded with, the leachate of 100 mL was recirculated per day and another more 100 mL one was collected and filtered for parameters analysis such as pH, conductivity, alkalinity and chemical oxygen demand (COD) etc. In addition, the gas production rate was recorded every day to measure the bacterial activity in the MSW biodegradation. From the results, it showed that 10 and 20 g l^{-1} fly ash added (g ashes addition per liter MSW ratios) bioreactors and 100 g l^{-1} bottom ash added bioreactors were found to enhance the gas production rate and the soluble concentration of alkali metals such as Ca, Mg, K, and Na as compared to the control one. The six soluble heavy metals of Cd, Cr, Cu, Pb, Ni and Zn in leachate were also found to be under inhibitory concentration for anaerobic digestion. Other trace metals such as Co and Mo etc were assumed to serve as the stimulatory micronutrients rather than to exert inhibitory effects on the microorganisms in the MSW anaerobic digestion.

In: Solid Waste: New Research
Editors: A. S. Martin and J. D. Suarez

SBN: 978-1-60456-809-7

Chapter 1

WASTE TO ENERGY: ON THERMAL TREATMENT AND ENERGETIC UTILIZATION OF SOLID WASTES

A. Karagiannidis[1], B. Bilitewski[2], G. Tchobanoglous[3], N. J. Themelis[4], M. Wittmaier[5] and Th. Tsatsarelis[1]

[1]Laboratory of Heat Transfer and Environmental Engineering, Department of Mechanical Engineering, Aristotle University, Box 483, 54124, Thessaloniki, Greece

[2]Technical University Dresden, Institute for Waste Management and Contaminated Site Treatment, Pratzschwitzer Str.15, 01796 Pirna-Copitz, Germany

[3]Department of Civil and Environmental Engineering, 1 Shields Avenue, University of California, Davis, CA 95616, USA

[4]Earth Engineering Center and Department of Earth and Environmental Engineering, Columbia University, New York, NY 10027, USA

[5]Institute for Circular Economy, Bremen University of Applied Sciences, Germany

ABSTRACT

Waste to energy has a clear dual role in the context of sustainable waste management: From the one hand, it has to reduce the volume and hazardousness of wastes that can receive no further treatment prior to disposal; on the other, it has to make the most out of the energy content of the treated wastes, in order to improve the economics and environmental performance of the whole process. This paper presents an overview of the state-of-the-art in this field, highlighting some related trends, needs, barriers, opportunities, drivers, pros and contras. The surveyed technologies fall under the following categorization:

- Conventional incineration of mixed wastes and residues with energy recovery and flue-gas treatment.
- Pyrolysis and gasification of mixed wastes, pre-treated fractions and residues.
- Co-firing of waste derived fuels in industrial furnaces.

- Anaerobic digestion of organic substrates and mixed wastes/residues in digesters.
- Thermal inertization / detoxification of hazardous waste fractions.
- Pretreatment of wastes for subsequent thermal treatment and energy recovery; waste derived fuels (refuse derived fuel, tire derived fuel, packaging derived fuel, solid recovered fuel).

ACRONYMS

BOF	Biodegradable Organic Fraction
BUWAL	Swiss Federal Office for Environment, Forest and Landscape.
BVS	Biodegradable Volatile Solids
CFB	Circulating Fluidized Bed
ESP	Electrostatic Precipitator
IC	Internal Combustion
MACT	Maximum Available Control Technology
MBT	Mechanical-Biological Treatment
MSW	Municipal Solid Waste
NIMBY	Not In My Back Yard
NIR	Near Infra Red
NYC-DOS	New York City Department of Sanitation
PM	Particulate Matter
PRF	Process Refuse Fuel
RDF	Refuse Derived Fuel
RVS	Refractory Volatile Solids
SDA	Spray Dry Absorber
SRF	Solid Recovered Fuel
TEF	Toxic Equivalent Factor
TEQ	Toxic Equivalent Quantity
TOC	Total Organic Carbon
TS	Total Solids
UBA	Umweltbundesamt
US-EPA	United States Environmental Protection Agency
VS	Volatile Solids
WTE	Waste to Energy

SECTION 1: ENERGY RECOVERY FROM MUNICIPAL SOLID WASTES

1.1. Thermal Energy in Municipal Solid Waste Components

Recoverable energy is stored in chemical form in all Municipal Solid Waste (MSW) materials that contain hydrocarbons; this includes everything except metals, glasses, and other inorganic materials (ceramics, plaster, etc.). By combusting such wastes, electricity and steam can be generated.

On the basis of the composition data for the "typical" MSW by Tchobanoglous (1993), and the atomic weights of the respective elements, Themelis and Kim (2002) showed that the chemical formula $C_6H_{10}O_4$ approximated the mix of organic wastes in MSW most closely. This formula corresponds to that of at least ten organic compounds, such as adipic acid and ethylene glycol diacetate. The heat of formation of most of these compounds is about -960 kJ mol^{-1} (Roinen, 1999). Representing the dry stream by the $C_6H_{10}O_4$ formula results in the following combustion equation:

$$C_6H_{10}O_4 + 6.5O_2 = 6CO_2 + 5H_2O \quad (1)$$

This reaction is highly exothermic and at the combustion temperature of 1000°C generates about 2700 kJ mol^{-1}. Since the molecular weight of $C_6H_{10}O_4$ is 146, the "theoretical" heat of reaction (i.e. in the absence of inerts or moisture) per unit mass of MSW is calculated to be 18500 kJ kg^{-1}. Similarly, if the MSW combustibles are simulated by the less oxidised compound $C_6H_{10}O_3$, the combustion reaction is:

$$C_6H_{10}O_3 + 7O_2 = 6CO_2 + 5H_2O \quad (2)$$

and generates 23000 kJ kg^{-1}.

1.2. Effect of Moisture and Inert Materials on Available Heating Value

The inclusion of moisture and non-combustible materials in the MSW affects the available heat in Waste-to-Energy (WTE) plants. To quantify these effects, it is assumed that the WTE plant provides steam to a standard power plant and that the exhaust gases leave the boiler at 120 °C and 0.135 MPa. Accordingly, the "waste heat"per kg of water in the feed is calculated at 2636 kJ kg^{-1}. The noncombustible materials in the feed, mainly glass and metals, end up mostly in the bottom ash. If it is assumed that the ash exits the combustion chamber at about 700 °C, the corresponding heat loss to inorganic materials fed with the combustibles is estimated to be as follows:

- Glass and other siliceous materials: 628 kJ kg^{-1}
- Iron: 420 kJ kg^{-1}
- Aluminium: 1134 kJ kg^{-1}

Since the iron/aluminum ratio in MSW is about 4 (for New York wastes - Table 1.1), the mean heat loss per kg of metal is estimated to be 544 kJ kg^{-1} (234 BTU lb^{-1}). Accordingly, the effects of non-combustibles on the heating value of RDF can be expressed as follows:

$$\text{Heating value of mixed MSW} = (\text{heating value of combustibles})*X_{comb} - (\text{heat loss due to water in feed})*XH_2O - (\text{heat loss due to glass in feed})*X_{glass} - (\text{heat loss due to metal in feed})*X_{metal} \quad (3)$$

and substituting numerical values for the heat of reaction and water and inorganic heat losses:

$$\text{Heating value of mixed MSW} = 18500X_{comb} - 2636XH_2O - 628X_{glass} - 544X\text{metal kJ kg}^{-1} \quad (4)$$

where X_{comb}, XH_2O, etc. are the fractions of combustible matter, water, etc. in the MSW.

Table 1.1. Detailed characterisation of New York City MSW (adapted from: SCS Engineers, 1992)

Waste component	% weight
Paper (analysis below:)	31.3
Corrugated cardboard:	4.7
Newspapers:	9.2
All other paper:	17.4
Plastics (analysis below:)	8.8
HDPE (clear and colour):	1.1
Films and bags:	4.8
PET:	0.5
Polypropylene, polystyrene:	0.9
PVC:	0.1
All other plastics:	1.4
Wood	2.2
Textiles	4.7
Rubber and leather	0.2
Fines	2.3
Other combustibles	2.3
Food waste	12.7
Grass/leaves	3.4
Brush/prunings/stumps	0.7
Disposable diapers	3.4
Miscellaneous organics	7.8
Glass (analysis below:)	5.0
Clear glass containers:	2.9
All other glass:	2.1
Aluminium	0.9
Ferrous metal	3.9
Hazardous waste	0.4
Bulky Items (appliances, furniture, etc)	5.0

It is preferable to separate the "wet", putrescible materials that contain a lot of water and also complicate handling and sorting of the rest of the waste, from the non-combustible fraction (metal, glass, other inorganic materials). Separating the noncombustible fraction also increases the thermal efficiency and ease of operation of the combustion plant. However, this increases the cost of collection and requires a market for the compost product.

The experimentally determined heating value of the "dry" stream, after separating the "wet" and the noncombustible fractions, is about the same as the thermochemical value calculated on the basis of Equation 1. Also, it is in the range of lignitic and sub-bituminous coals that are still used in many power plants. Therefore, separating and using the "dry combustible" MSW as a fuel can reduce, by as much as ton ton^{-1}, the need for mining coal. It should be noted that a typical largescale power plant uses approximately 11000 kJ to produce 1 kWh; therefore, combusting one ton of the "dry" and "wet" mix should generate about 1100 kWh. However, due to low thermal efficiencies, the best WTE plants produce a gross of 710 kWh ton^{-1} of mixed MSW (EAC, 1999). Obviously, there is room for improvement. Figure 1 shows the effect of moisture on the heating value of MSW. The bold line represents the heating value calculated from Equations 1 and 4 (assuming that the composition of organic matter in MSW can be represented by the molecular formula $C_6H_{10}O_4$). It can be seen that this line is fairly representative of the heating value of several waste materials reported in the literature. Of course, wastes consisting mostly of plastics cannot be represented by $C_6H_{10}O_4$ but by lower oxygen organic materials that have higher heating values. The opposite is true for wastes that contain mostly paper where cellulose ($C_6H_{10}O_5$) is the predominant compound. Figure 1.1 shows the projected heating values for various $C_6H_{10}O_x$ compounds as a function of moisture content.

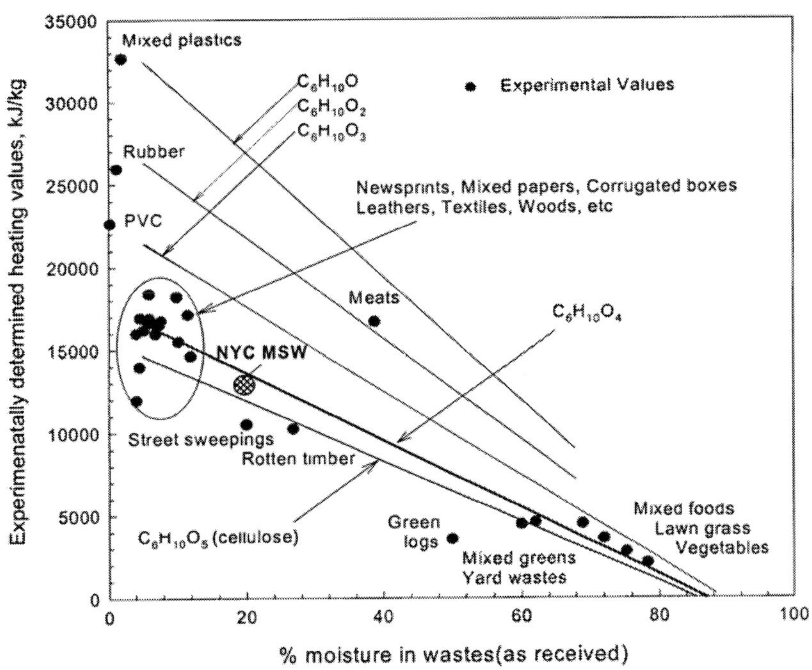

Figure 1.1. Effect of moisture on heating value of MSW materials (points shown are experimental values; solid lines show thermochemical calculations for various organic compounds).

1.3. Solid Waste as a Renewable Source of Energy

In the traditional sense, renewable sources of energy are those that nature can replenish, such as waterpower, windpower, solar radiation and biomass (wood and plant waste). However, MSW contain a large fraction of paper, food wastes, cotton and leather, all of which are renewable materials under proper stewardship of the Earth. Municipal solid wastes also contain man-made plastics, rubber and fabrics that were produced using non-renewable fossil fuels. All these materials were produced because they were needed by humanity. Although it is desirable to minimize the amount of materials used per capita, and also the generation of wastes during production and distribution of goods, a certain quantity of wastes will always be generated. Therefore, if the produced waste is to be disposed in some way, there is a continuous new stream to replenish it. In 2001, the US EPA reported a MSW generation of 229.2 million short tons (US EPA, 2003). Even if one uses a more stringent definition of the term "renewable", one that includes only material from non-fossil sources, about 64% of the US MSW after material recovery for recycling plus composting are derived from renewable sources (Table 1.2).

Table 1.2. Reported Composition of US MSW in 2001, before and after recycling and composting (US EPA, 2003)

Material	Total MSW (million tons)	Recovered (million tons)	Not recovered (million tons)	% of MSW after recovery
Paper and board	81.9	36.7	45.2	28.0
Yard trimmings	28.0	15.8	12.2	7.6
Food scraps	26.2	0.7	25.5	15.8
Wood	13.2	1.3	11.9	7.4
Textiles	9.8	1.4	8.4	5.2
Sum of renewable materials	*159.0*	*55.9*	*103.1*	*63.9*
Rubber and leather	6.5	1.1	5.4	3.3
Plastics	25.4	1.4	24	14.9
Glass	12.6	2.4	10.2	6.3
Metals	18.1	6.3	11.8	7.3
Other	7.7	0.9	6.8	4.2
Total	229.2	68.0	161.2	100.0%

Note: All tonnages are reported in short tons (1.1 short tons=1 metric ton).

This fraction of MSW could be used as clean, sustainable and arguably renewable fuel for the production of electricity and steam. The remaining non-renewable portion, however, has to be either separated or accepted as part of the fuel. BioCycle's annual garbage survey reported the generation of about 370 million tons of MSW in the US for 2002 (i.e., nearly 50% higher than the EPA report for 2001 (US EPA, 2003)), of which about 28.5 million short tons (or 7.7%) were processed for energy recovery in WTE facilities (Kaufman et al., 2004).

1.4. The Renewable Status of Waste-To-Energy

A detailed review of the opposition to renewable energy status for WTE facilities in the U.S. has shown that there are five arguments repeated in a variety of statements:

1. There should be much less municipal solid waste generated and what is generated should be recycled. When this is done, there will not be waste fuel for waste-to-energy power plants in the future. Therefore, waste-to-energy is not a renewable source of energy.
2. Granting renewable status to waste-to-energy will inhibit efforts to recycle wastes.
3. In considering renewable energy status, WTE should not be credited with the avoided environmental impacts of landfilling.
4. WTE is incineration and, on a per kWh basis, the generation of electricity by WTE is more polluting than by coal-fired power plants.
5. Granting renewable energy status to WTE will lead to building new WTE facilities in low-income areas that are already suffering environmental injustice.

The reasoning behind these five arguments runs contrary based to the known facts. These facts are discussed below.

Con: After successful implementation of waste reduction and recycling concepts WTE facilities will run out of MSW fuel.

As noted earlier, the U.S. generated about 370 million tons of MSW in 2002. The nationwide recycling rate was 26.7%. The rest was either combusted in WTE plants (7.7%) or landfilled (65.6%) (Kaufman et al., 2004). It cannot be expected that the best possible practice of waste reduction and recycling will lead to the often-acclaimed "zero waste" scenario.

For example, intense effort by the New York City Department of Sanitation (NYC-DOS) in the 1990s to increase recycling led to a maximum annual rate of 700,000 short tons (total collected paper, metal glass, and plastic) or 16% of the 4.5 million tons of total MSW collected by NYC-DOS. However, a study by the Earth Engineering Center showed that of the amount recycled, only 550,000 tons were actually recycled and the non-marketable residues (mostly glass and plastics) were disposed in landfills (Themelis and Todd, 2004). New York City is not unique in this regard: the US EPA reported that in the year 2001, of the 25.4 million tons of plastics discarded in MSW, only 1.4 million tons were recycled; also of the 159 million tons of combined renewable materials (paper, yard trimming, food scraps, textiles and wood) only 56 million tons, i.e. 35% of the total, were recycled (US EPA, 2003; compare also Table 1.2). It is a sad fact of life that at least these fifty million tons of combustible materials end up each year in landfills, instead of being used to reduce the mining of coal or the importation of fuel oil.

The difficulties in recycling most of the MSW are illustrated by the Japanese experience, a nation where waste minimization and management are a top priority of the central government. Although Japan has been a world leader for several decades in waste reduction and recycling systems, 40.3 million metric tons of the 53.7 million metric tons of generated MSW were combusted in 1999 (Japanese Ministry of Environment, 2000).

The European Union (EU) has put in place many intensive waste reduction and recycling programs. Nevertheless, the EU is a big user of WTE (40 million metric tons of municipal solid wastes are presently combusted in WTE facilities) and the EU Landfill Directive (1999/31) requires that, as of 2020, only non-recyclable and non-combustible wastes be disposed in landfills (European Union, 1999).

Con: Granting renewable status to WTE will inhibit efforts to recycle wastes

Recycling and WTE are not competing but complementary and essential means of waste management. WTE does not displace recycling; it displaces landfilling of trash. In fact, the recycling rate for communities with WTE facilities was reported to be 33% (Kiser, 2003), substantially higher than the average U.S. rate of 26.7% (Kaufman et al., 2004). At most WTE plants, materials such as ferrous metals are recovered from the waste stream either before or after combustion and thus, diverted from landfills.

The comparison of WTE to recycling, rather than to landfilling, is misleading to say the least: Obviously, WTE plants do not choose to combust recyclable materials. They have to process all the mixed waste that is delivered to them: plastic wastes, paper wastes, textile wastes, wood wastes, food wastes, disposable diapers and all other wastes. The 28.5 million short tons of MSW that are combusted annually in the US, as well as the hundreds of millions of tons landfilled, represent the real world of used materials that cannot be recycled.

Con: In deciding renewable status, WTE should not be credited with the avoided environmental impacts of landfilling.

WTE is the only form of renewable energy that: a) reduces environmental impacts of fossil fuel use; and also, b) reduces the environmental impacts of the only other alternative for waste disposal, landfilling. Therefore, it is eminently fair to consider both of these advantages. The search for renewable energy sources is motivated by the desire to reduce use of fossil fuels. WTE facilities in the US produce about 500 kWh per ton MSW, thus obviating the use of either 0.3 tons of coal or 1 barrel of fuel oil per ton of MSW combusted. In addition, WTE reduces the need for landfilling by 1 ton of municipal solid waste per ton of trash combusted. Thus, the generation of electricity by means of WTE provides greater environmental benefits than any other source of renewable energy, for the simple reason that in addition to the benefit of reducing the use of fossil fuels, as all other renewable energy sources do, it also avoids the environmental impacts of landfilling.

The primary reason that numerous communities across the US have spent billions of dollars to combust rather than landfill MSW was to avoid the environmental impacts of landfilling, particularly the potential of groundwater contamination, most feared in sandy soils, such as Florida and Long Island, New York. These communities did not do so to avoid recycling but to avoid landfilling, despite the fact that WTE facilities are much more expensive to build and operate than modern landfills. Of course, a community's decision, and economic ability, to proceed is influenced by the proven fact that WTE, in addition to avoiding landfilling, generates 500 kWh per ton of MSW combusted. This provides an economic incentive gained from the generation of electricity that in the US currently ranges from $15 to $35 per ton of solid waste.

In sum, the objective of renewable status legislation is to provide an economic incentive that encourages the development of alternative energy resources in order to reduce the environmental impacts resulting from the extraction and combustion of fossil fuels. WTE provides this environmental advantage, because, on the average, the combustion of one ton of municipal solid waste produces electricity equivalent to 0.3 tons of coal or one barrel of oil. In addition, WTE is the only renewable energy source to offer an additional environmental advantage: the avoidance of the environmental impacts of landfilling MSW.

Con: WTE is incineration and the generation of electricity by waste-to-energy is more polluting than coal-fired power plants, on a per kWh basis

There are nearly one hundred WTE power plants in the US and more than 1600 incinerators. Incinerators were developed to control the spreading of diseases and limit the number of vectors that are attracted by waste. WTE facilities are the result of the technological evolution of incinerators over several decades and they are equipped after the implementation of the EPA mandated Maximum Available Control Technology (MACT) regulations with gas control facilities that are superior to any type of incineration and, also, to many coal-fired power plants in the U.S. Also, in contrast to coal-fired power plants, WTE serve two purposes, conserving (a) non-renewable fossil fuels and also (b) land for landfilling.

The desirability of fossil fuels is based on their heating value, hydrogen content, emissions of minor constituents and energy generated. For these reasons, natural gas is preferable to fuel oil and fuel oil to coal. It is obvious that MSW is at the very bottom of the list of desirable fuels, since it is a very nonhomogeneous material consisting of all the solid products that citizens use, cannot or do not recycle and discard. The heating value of MSW is less than one half that of coal (although nearly as much or even slightly higher to that of low quality coal, e.g. lignite used in many countries where locally available like e.g. Greece) and the heat-to-electricity conversion efficiency of WTE facilities is two thirds that of coal-fired power plants. Therefore, a comparison of emissions from WTE and fossil-fuel power plants, on the basis of kWh's of electricity generated, serves no purpose but to show the known fact that MSW is not as good a fuel as coal, or oil.

In real world terms, the environmental impact of emissions is a function of the product of contaminant concentration per cubic meter of gas emitted and the volume of gas emitted from a particular source. As noted earlier, after implementing the MACT regulations, the US WTE industry decreased its mercury and cadmium emissions in the gas emitted by a factor of nearly one hundred. Dioxin emissions were reduced by a factor of nearly one thousand (Themelis, 2003). Yet, this enormous improvement in emissions control that cost more than one billion dollars is being disregarded by those who now oppose WTE as vehemently as they did in the eighties. As a result of MACT, the US WTE facilities now emit a small fraction of the mercury and cadmium emitted by the US coal-fired power plants (Figure 1.2).

In addition, the total volume of flue gas emitted by coal-fired power plants that provide 51% of the US electricity amounts to about 2,400 billion cubic meters per annum versus about 180 billion cubic meters from all US WTE power plants. The overall result is that US coal-fired power plants emit a total of about 40 tons of mercury to the atmosphere while mercury emissions from US WTE plants are about 2 tons.

The average mercury concentration in US MSW has been estimated at about one part per million (Themelis and Gregory, 2002). On this basis, the amount of mercury disposed annually in New York landfills is about 13,000 kg per year. Most of the mercury in MSW is in metallic form (fluorescent lamps, thermometers, etc.). The vapor pressure of mercury at landfill temperatures (40°C) is 0.007 mm Hg, as compared to the vapor pressure of water of

5.67 mm Hg at 40°C. Therefore, if an exposed water droplet evaporates in one hour, then a mercury droplet of the same size will evaporate in four weeks. Also, the conditions in an MSW landfill (elevated temperature, moisture, and reducing capacity) are favorable for aqueous mobilization of mercury (e.g. in the form of methyl mercury). However, since both gaseous emissions and aqueous mobilization from landfills are dispersed over hundreds of acres, in contrast to the point sources of WTE stacks, they are difficult to measure. Mercury emissions from landfills and also dioxin emissions from flared landfill gas are now the subject of research internationally.

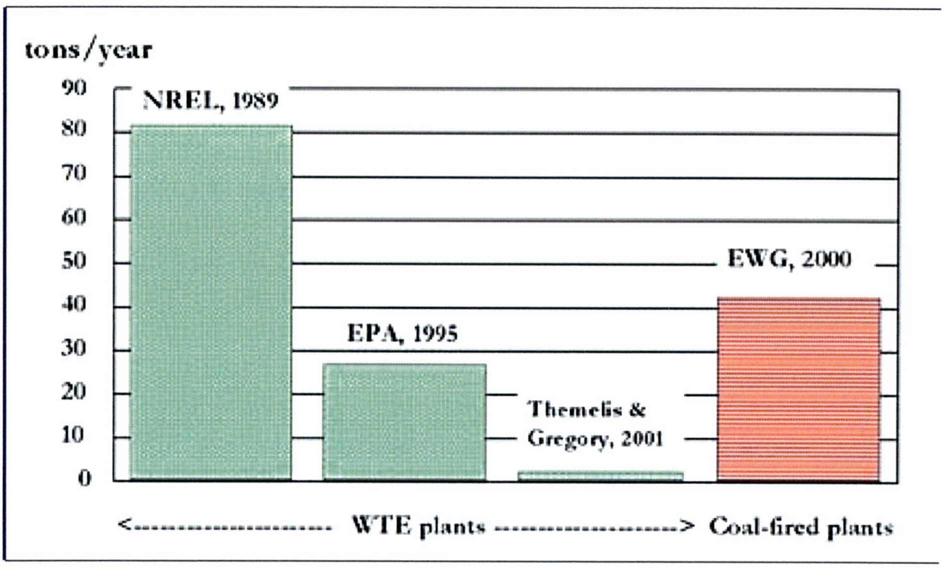

Figure 1.2. Mercury emissions from WTE (1989-1999) and coal-fired power plants (Themelis and Gregory, 2002).

Con: Granting renewable energy status to WTE will lead to building new WTE facilities in low-income areas that are already suffering environmental injustice.

WTE facilities, as well as any other industrial installations, should be built at locations where they will improve, rather than deteriorate, the quality of life of the surrounding community, in terms of esthetics, provision of jobs and services, as well as environmental quality. While it is very difficult to site new WTE facilities, there is growing public acceptance of already built WTE plants. Initially, host communities respond with the NIMBY (Not in my backyard!) syndrome. After a certain period of time, NIMBY may translate into proactive movement through civil participation and open communication between all parties involved (Scharff, 2003).

For example, at this time, New York City exports millions of tons of MSW to other states for landfilling. Although it is not necessary to build WTE facilities for New York City within the city (New York City already combusts 0.5 million tons of MSW in a New Jersey WTE facility), new WTE facilities can be used to improve, physically, esthetically and environmentally, some brownfield areas of the city. Opponents of granting renewable status to WTE facilities in New York State have claimed that this will lead to building these plants in places where it will result in "environmental injustice" in New York City. Yet, throughout the nation and the world, there are hundreds of WTE plants that have not been accused of creating environmental injustice and in fact are welcome by the host communities.

It is true that, in the past, New York City has not been "just" in locating waste facilities in some boroughs (e.g. one can compare the state-of-the-art North River waste water treatment plant near Columbia University with the Hunts Point wastewater treatment plant; also the multi-decade imposition of Fresh Kills on the citizens of Staten Island). However, the mistakes of the past need not to be repeated in the future. For instance, a properly designed WTE plant could be accompanied by refurbishing of nearby rundown facilities or by the reclamation of industrial brownfields, creating an industrial "ecopark" where the electricity and waste heat of the WTE plant is used for heating and cooling adjacent industries[1]. Such a municipal project would result in closing a number of nearby transfer stations, some of which resemble temporary landfills, stopping the transport of millions of tons of MSW over already crowded bridges and hundreds of kilometers of throughways and country roads, and landfilling the New York City waste in the green fields of other states. One of many existing example of avoiding environmental justice is the case of the WTE plant in Brescia, Italy, that was built on a former landfill. Emissions of pollutants have significantly decreased throughout the City after the WTE facility started to provide district heating and the citizens of Brescia are rightly proud of their modern WTE (Bonomo, 2003).

1.5. Emission Comparison of Waste to Energy to Coal-Fired Facilities

The conversion of MSW in WTE facilities has been recognized globally as a means to conserve fossil fuels and improve environmental quality by lessening the amount of waste to be landfilled (Themelis, 2003). Worldwide, over 130 million tons of MSW are combusted annually in WTE facilities that produce electricity and steam for district heating. In the US alone, about 30 million tons of MSW are combusted in WTE plants to produce about 2.8 GW of electricity and some steam for district heating (Integrated Waste Services Association), which is approximately 0.3 % of total US electricity production. However, one of the major concerns in combustion of MSW in WTE plants is the emission to the atmosphere. The WTE emissions of concern are trace organic compounds, particularly polychlorinated dioxins and furans, volatile trace metals such as mercury, lead and cadmium, total particulate matter, and acidic gases such as hydrogen chloride, hydrogen fluoride, sulfur dioxide and nitrogen oxides.

Of the one billion tons of coal produced annually in the US, approximately 90% is burned to generate electricity. In 2000, 52% of the 3,800 billion kWh of the net power generated by the electric power industry was from coal-fired power plants (Energy Information Administration, 2000). Coal-fired generating units produce more sulfur dioxide and nitrogen oxides than oil or gas-fired units as coal generally contains more sulfur than other fossil fuels, in addition to being the principal fuel in the US. As part of the 1990 Clean Air Act mandates, US-EPA promulgated in 1995 new air pollution control standards for large WTE facilities (MSW combustors). These standards required facilities to implement the Maximum Achievable Control Technology (MACT). The 1995 MACT standards apply to facilities with large units, i.e. that combust more than 250 tons each day of MSW. MACT standards for

[1] The most probable future scenario for the 1st WTE facility in Greece (the only EU country without any WTE facility by the year 2007) is such, namely for an industrial park of the Greater Thessaloniki area.

small units that combust less than 250 tons of MSW per day were proposed in the summer of 1999. Table 1.3 shows the existing MACT rules for existing large units.

Table 1.3. MACT rules for existing large MSW combustion units

Emissions	Limits	
Dioxin/Furans		
Large-ESP Equip Units	60 ng/dscm	
All others	30 ng/dscm	
Metals		
Cadmium	0.04 mg/dscm	
Lead	0.44 mg/dscm	
Mercury	0.080 mg/dscm or 85% removal	
Acid Gases		
Sulfur Dioxide	29 ppm or 75 % removal	
Hydrogen Chloride	29 ppm or 95 % removal	
Nitrogen Oxides	Option A	Option B
Mass Burn/Water Wall	205 ppm	180 ppm
RDF	250 ppm	220 ppm
Mass Burn/Rotary	250 ppm	220 ppm
Fluidized bed	180 ppm	160 ppm
Particulates	27 mg	
Fugitives	Visible less than 5% of time	

Source: http://www.wte.org.

1.5.1. Emissions

The acid gases SO_2, SO_3, HCl and HF are all removed by the same kind of control equipment, i.e. dry or wet scrubbers. The design of scrubbers for WTE systems is largely based on experience obtained from coal-fired power plants. Mass transfers and chemical reaction rates are similar in coal-fired combustion and MSW combustion. However; MSW combustion has significantly lower levels of sulfur oxides, higher levels of HCl and lower levels of fly ash compared with that of coal combustion. Since HCl and HF are easier to scrub than SO_2, a system designed to scrub SO_2 will have no problem with these gases with respect to removal efficiency. Wet limestone scrubbers are the most commonly used control devices for large coal-fired plants. A more recent development is the dry scrubbing system, which may use either dry sorbent or an aqueous sorbent solution (spray dry absorber, SDA) that is sprayed in the flue gas. Dry systems are generally less expensive to install than wet scrubbers but have higher operating costs because larger amounts of sorbents are required (Ruth, 1998).

1.5.1.1. Sulfur Oxides

Gaseous sulfur oxides from coal combustion are primarily sulfur dioxide (SO_2), plus a much smaller quantity of sulfur trioxide (SO_3) and gaseous sulfates. These compounds form as the organic and pyritic sulfur in coal are oxidized during the combustion process. On the average, about 95% of the sulfur present in bituminous coal is emitted as gaseous SO_x whereas somewhat less of sulfur is volatilized when sub-bituminous coal is fired (US-EPA, 1993).

1.5.1.2. Nitrogen Oxides

Nitrogen oxide () is formed in MSW combustion facilities in two separate pathways: thermal fixation of molecular nitrogen present in the combustion air (thermal NO_x) and the conversion of nitrogen present in the fuel (fuel NO_x). While thermal fixation is extremely sensitive to temperatures above 1800 K, formation of fuel NO_x is insensitive to temperature and depends primarily on the availability of oxygen. Conditions that promote good mixing of fuel and air tend to increase the conversion of nitrogen in the fuel to NO_x and raise emission levels (Ruth,1998).

NO_x emissions from coal combustion are primarily nitric oxide (NO) and a small quantity of nitrogen dioxide (NO_2) by volume. Nitrous oxide (N_2O) is also emitted at a few parts per million. As in the case of MSW combustion, NO_x is formed from thermal fixation of atmospheric nitrogen in the combustion flame and also from oxidation of nitrogen in the coal. Bituminous and sub-bituminous coal usually contain from 0.5 to 2% nitrogen by weight; therefore, fuel nitrogen can account for up to 80% of the total nitrogen oxides from coal combustion.

1.5.1.3. Particulates

Particulate matter (PM) composition and emission levels are a complex function of boiler firing configuration, boiler operation, pollution control equipment and coal properties (US-EPA, 1993). In pulverized coal systems, combustion is almost complete; thus, the emitted PM is primarily composed of inorganic ash residue. Electrostatic precipitators (ESPs) and fabric filters are used to control particulate emissions from coal-fired power plants. Overall particulate removal efficiencies are often well over 99% and since most trace elements are associated with particulates, most trace elements are collected along with the particulates. New WTE plants and the majority of existing large WTE plants use a combination of fabric filters and dry scrubbers.

1.5.1.4. Trace Metals

Trace metals are emitted both during coal and MSW combustion. The quantity of any given metal emitted in general depends on: physical and chemical properties of the metal itself; concentration of the metal in the fuel, type of the particulate control device used, and its collection efficiency as function of particle size. In addition, emission of trace metals from the combustion of fuel depends not only on the element concentrations in the fuel itself, but even more strongly on the chemical and physical phenomena associated with the combustion process. Some metals become concentrated in certain particle streams from combustor (e.g., bottom ash, collector ash, and flue gas particulate) while others do not. Various models have been developed to describe this partitioning behavior (Klein et al., 1975). The phenomena involved in partitioning are important because they determine how and where trace elements are emitted and impact the type of emission control device that may be required. The same principles that apply to the partitioning of trace metals in coal combustion may also apply to the combustion of MSW.

1.5.1.5. Cadmium

Cadmium oxide is the predominant form of cadmium in the atmosphere. However, the toxicology of cadmium does not appear to be dependent on the chemical forms. Because a typical residence time of particulate cadmium is between 1 and 10 days, long-range transport (up to thousand km) can occur. Emission control measures for both coal-fired and MSW combustion include better combustor design and operating practices that are directed at nitrogen oxides and particulate matter and add-on air pollution devices that are designed for acid gas and PM control.

1.5.1.6. Mercury

Mercury is emitted in several chemical forms or oxidation states, including elemental and oxidized mercury, which impact not only in its removal but also on its transport and subsequent deposition once released into the atmosphere (Ruth, 1998). Mercury is probably the trace element of most environmental concern, as it tends to bio-accumulate in the food chain once released into the environment. Mercury is primarily emitted as a vapor and was not effectively controlled by the pollution control equipment used during most of the 20th century. However, activated carbon injection in the flue gas results in the adherence of mercury molecules on the carbon particles which are then recovered in fabric bags through which the gas is filtered. This is the control system used in WTE plants that have implemented the MACT regulations.

1.5.1.7. Dioxin/Furans

The chlorinated dibenzo-p-dioxins and chlorinated dibenzofurans are formed as unwanted byproducts of certain chemical processes during the manufacture of chlorinated intermediates and in the combustion of chlorinated materials.

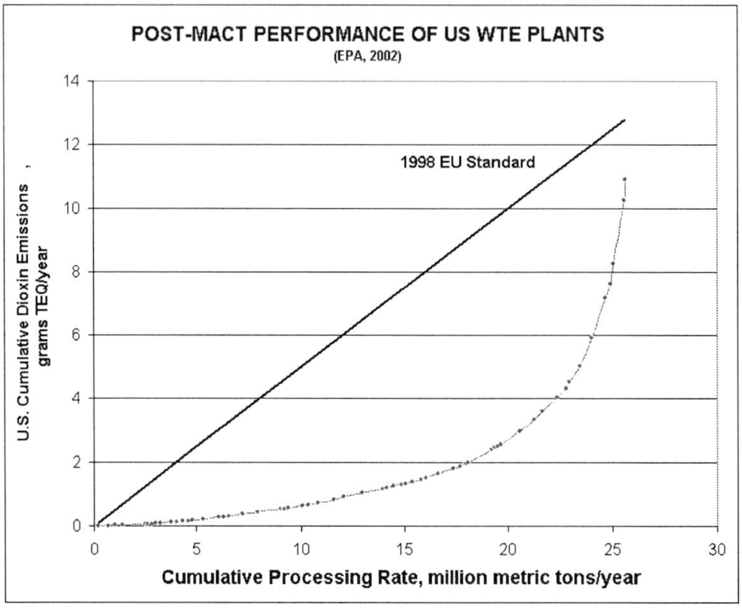

Figure 1.3. Cumulative dioxin emissions (in grams TEQ) of U.S. WTE facilities in 2000 (each dot represent a WTE facility).

Dioxin and furans are emitted into the atmosphere from a wide variety of processes such as waste combustion, fossil fuel combustion, iron and steel industries/scrap melting, combustion aided-metal recovery, internal combustion engines, carbon regeneration, organic chemical manufacture and use, forest fires and cement manufacture (US-EPA, 1997). The toxic equivalent factor (TEF) was established to relate the toxicity of each compound on a 0-1 scale, where 1 is the most toxic dioxin compound known. On the basis of these factors and the prevailing distribution of the different types of dioxin and furans in the flue gas of WTE facilities, it has been established that the prevailing ratio of total toxic dioxin to toxic equivalence is approximately 50. Which means, 50 grams of total dioxin and furan emission corresponds to about 1 gram of toxic equivalent (TEQ). The dotted line of Figure 1.3 (Deriziotis and Themelis, 2003) shows the post-MACT cumulative emissions of dioxins of the US WTE facilities. The diagonal straight line represents the allowable limit of toxic dioxins (grams TEQ) using the present EU limit of 0.1 nanogram per cubic centimeter and the cumulative processing rate of MSW.

1.5.2. Conclusions

The estimates showed that, per unit of thermal energy input and also per unit of power generated, the emissions of sulfur dioxide, particulate matter and nitrogen oxides were lower in WTE facilities than for coal-fired power plants; on the same basis, hydrogen chloride emissions are 5 times higher in WTE as compared to coal-fired utilities and emissions of cadmium, lead and mercury from WTE facilities were comparable to that of coal-fired plants. With regard to dioxins/furans, the emission factors of WTE are considerably lower than for coal-fired utilities. Table 1.4 shows emission factors from coal-fired plants and WTE in the US.

Table 1.4. Summary of emission factors of coal-fired and WTE plants in the USA

Emissions (Year 2000)	Emission Factors			
	In terms of thermal energy input: $g/10^6 J$		In terms of electrical energy output: g/Wh	
	Coal-fired	WTE	Coal-fired	WTE
Acids				
Sulfur Dioxide	0.45	0.01	4.9	0.21
Hydrogen Chloride	0.017	0.087	---	---
Nitrogen Oxides	0.19	0.15	2.1	2.4
Particulate Matter	0.027	0.002	0.29	0.037
Trace Metals	$g/10^{12}$ J		Lb/TWh (g/GWh)	
Cadmium	1.8-7.7	1.1		17
Lead	14-50	10		110
Mercury	0.2-12.9	7.2		0.11
Dioxin/Furans	60[a]	15[a]		

[a]in grams TEQ (grams Toxic Equivalent Quantity).

SECTION 2: MUNICIPAL SOLID WASTE INCINERATION

2.1. Overview in Europe

Over the last decade, several environmental legislations and landfill bans were put into effect. Due to these restrictions in landfilling, thermal treatment plays an important role in European waste management systems, whereas an increasing role can be expected in the future. In this context, the Waste Incineration Directive is fundamental, as all EU member states have to meet the same incineration standard from now on. Furthermore, it is necessary to take into account the different level of the waste management infrastructure in several European countries.

Table 2.1 illustrates the differences of the relevance of waste incineration in the European countries. While most of the highly industrialized countries of Western and Northern Europe (orange marked) incinerate more than 40 % of MSW the rate in the new member states (green marked) is less than 10 %. In Switzerland nearly the entire amount of MSW is being used in WTE plants.

Table 2.1. Amount of household waste (total and incinerated amount) in European countries

Country	MSW total [mill. ton/y]	MSW (incinerated) [mill. ton/y]	Rate of thermal treatment [%]
Austria	1.27	0.55	43.3
Belgium	2.15	1.04	48.4
Denmark	3.10	1.85	59.7
France	22.37	9.21	41.2
Germany	17.05	9.72	57.0
Netherlands	4.79	3.75	78.3
Norway	0.83	0.39	47.0
Sweden	2.35	1.94	82.6
Switzerland	2.58	2.50	96.9
Finland	1.40	0.08	5.7
Italy	24.09	1.78	7.4
Portugal	4.19	0.94	22.4
Spain	13.1	1.48	11.3
United Kingdom	24.61	2.64	10.7
Czech Republic	2.95	0.21	7.1
Hungary	4.18	0.35	8.4
Latvia	1.13	0.03	2.7
Poland	12.74	0.002	0.02
Slovakia	1.02	0.13	12.7

In Europe, a wide variety of WTE plant sizes and types exist at present. France for example has for the most part small installations (average capacity = 40 kt/a) whereas the Netherlands have the largest installations with an average capacity of 460 kt/a. The most commonly used technology for MSW incineration is the grate furnace, but other proven technologies exist as well (such as fluidized bed furnaces). Figure 2.1 shows the incineration plants in Europe and Figure 2.2 the growth in million ton/y capacities.

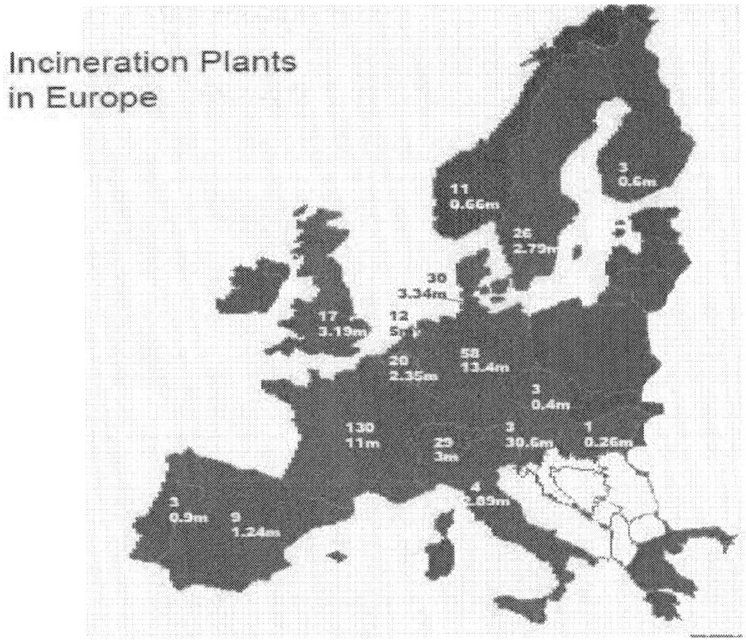

Figure 2.1. Distribution of incineration plants in Europe (numbers of facilities and millions of tons MSW incinerated annually) in the year 2003.

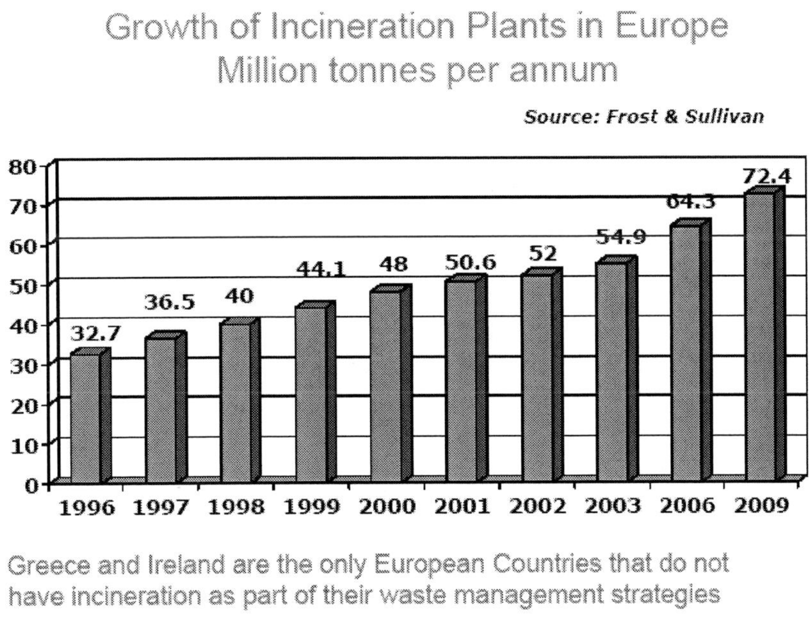

Figure 2.2. Rise of capacity of incinerator plants in Europe in million ton/y.

There were 67 MSW incinerators in operation in Germany at the end of 2005. The total capacity of these units is approximately 16,7 Million. ton/y. All wastes are processed in facilities with energy recovery and heat and/or electricity generation. 72 % of the total energy potential was used for heat generation (13609 GWh) and approximately 28 % for electricity generation (5227 GWh).

Beside the volume reduction and mineralization of waste, energy recovery has increasingly become an important objective. Due to the fact that waste has a role as a renewable energy source, it is attracting more and more attention in this respect.

2.2. Current Technologies

MSW incinerators are `at the end of pipe` and various waste categories with several characteristics have to be treated. This requires specifically designed combustors. On principle, a waste incineration plant can be divided in the main thermal process and the flue-gas cleaning. The thermal process contains first the storage, handling and waste feeding, followed by the combustion in the furnace, as well as the heat recovery with steam and electricity production.

The most common technology for the incineration of MSW is the grate system combined with a combustion chamber. In Germany, more than 90% of MSW incinerators are grate firing systems. These systems require minimal pre-processing and are implemented in facilities of varying grate size from 2 t/h to more than 40 t/h. The moving grates can be divided into:

- Reciprocating grates.
- Roller grates.
- Reversed feed grates.

A disadvantage of the grate systems is the large excess of air (λ=1.6-1.8) which leads to energy losses over the stack by the flue gas. Independent of the incineration system, the combustion temperature has to be in the range of 850-1100 °C. The lower limit guarantees the complete destruction of harmful organic chemicals and the upper limit should prevent from unacceptable high production of thermal NO_x (cf. Section 1). The residence time at minimum temperature of 850 °C for the gas resulting from the incineration has to be 2 seconds in the presence of at least 6% oxygen. The efficiency of MSW incinerators with electricity generation is around 20% and can go up to 28%. Compared to fossil fired plants these values are low and the main limiting factors are the low steam parameters of approximately 400°C and 40 bar.

The cost structure of a MSW incinerator depends on many factors like size and age of facility, kind of energy recovery, system of flue-gas cleaning and degree of utilization of slag and by-products of the flue-gas cleaning system. Therefore the treatment prices vary in a wide range between 120 and 280 €/T.

2.2.1. Mass-Burn Plants

More recently, "mass burn" incinerators were provided with emission control systems but many were not used for energy recovery. Today, mass burning of MSW with power generation, in WTE plants is practiced widely in the US and other developed countries and does not require pre-processing of waste, apart from manual removal of bulky items like "white goods". The rest of the waste is charged to the furnace by means of a mechanical "claw" that deposits full garbage bags and other items at the feed end of a metal grate that moves the waste materials slowly through the combustion chamber. Because of the large size of the items deposited on the grate mechanism, the rates of mass transfer and chemical reaction with the oxidizing gas are relatively slow. Because of the low rate of oxidation, a very large combustion chamber and grate are required and the intensity of combustion (rate of heat generation per unit volume) is correspondingly low. Therefore, the temperatures generated in the combustion chamber are in the order of 900°C and the ash does not reach the point of fusion or semi-fusion. There have been improvements over the years in combustion efficiency and pollution control of mass-burn WTE plants but the question has risen as to whether this technology has been surpassed by newer ones that are discussed in the following sections.

2.2.2. Fluidized-Bed Plants

Combusting of MSW in fluidized bed reactors is used extensively in Japan. This method requires shredding (down to 5 cm) and removing of inert materials like glass and metals from the feed to the fluidized bed reactor. The remainder is fed on top of a fluidized bed of sand or limestone. Combustion under these conditions is more efficient and results in even temperatures and higher energy recovery, lower amounts of non-oxidized materials leaving the combustion chamber, and less excess air than mass-burn plants. Fluidized-bed combustors operate at temperatures in the range of 830–910°C and can use additional fuel as required so that they can burn materials with very high moisture content. Because of the lower uniform temperatures, "slagging" and corrosion problems in the furnace are kept to a minimum. Fluidized beds also have a potential for solid agglomeration in the bed, if salts are present in feed, and may provide insufficient residence times for fine particles. The non-uniformity and fluctuation of heating values in the feed is compensated by the large total mass of inert solids in the bed.

2.2.3. A state-of-art US plant

The SEMASS facility at Rochester, Massachusetts is a good example of a state-of-art WTE plant. The feed material consists of the entire (i.e., wet+dry) MSW stream. The facility takes in waste by covered rail car and truck from 50 communities in a 110-km radius. The plant consists of three parallel combustion units and processes over one million tons per year. The first two units were built in 1989 and Unit 3 in 1994.

2.2.4. Waste Pre-Processing

Waste brought to plant is dumped on a tipping floor (Figure 2.3). The waste is loaded onto conveyors that pass inspectors who look for bulk waste that could jam the shredders, or for hazardous waste; these items constitute about 1.6% of the incoming material. The waste is then shredded in one of two large hammermill shredders that produce a blended material of 15-cm size coarse RDF. The shredded material is conveyed under overhead belt magnets for the first round of ferrous metal recovery and is then stored in bays in a closed building. This material is called Processed Refuse Fuel (PRF) and apparently can be stored in a storage building for long periods. Because some of the moisture has evaporated during shredding and storage, any moisture left is distributed and absorbed throughout the PRF; therefore the PRF does not have the acrid smell of raw waste and does not attract rats or flies.

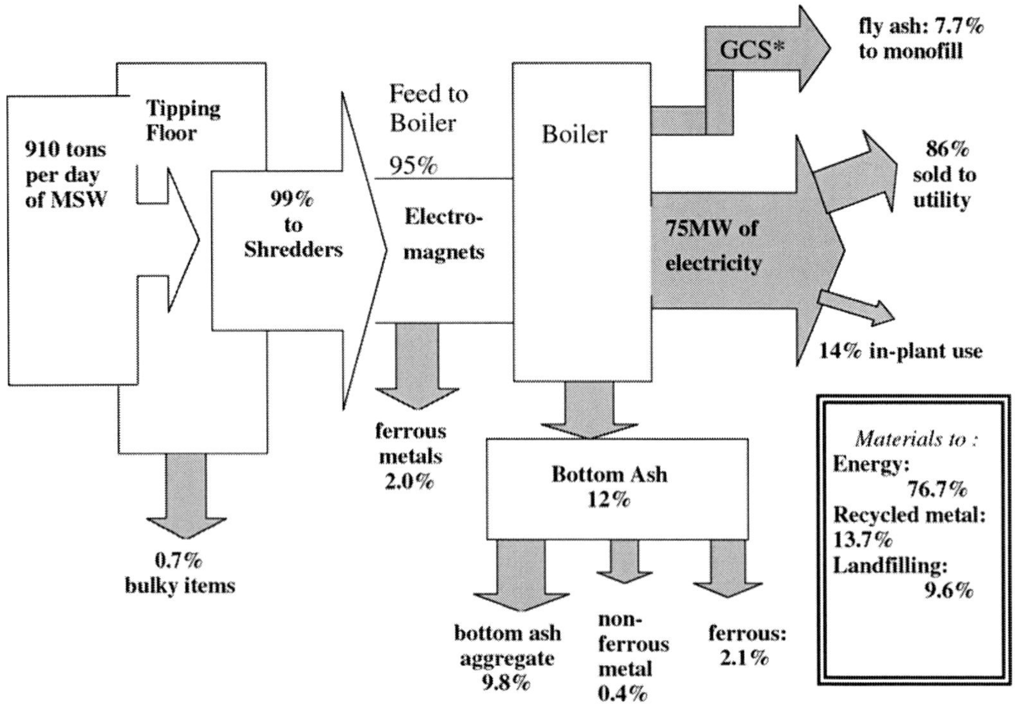

Figure 2.3. Materials flow sheet of one of three SEMASS WTE units, at Rochester, MA (*gas control system: dry scrubbing, activated carbon Injection, fabric filter bag-house).

2.2.5. Combustion Chamber

The PRF is conveyed to bins and from there is ejected through inclined chutes into the three combustion chambers. The feed rate of PRF into the boilers is adjusted by means of automated temperature controls. The PRF contains a lot of small, light particles that with the help of deflectors at the bottom of the chutes and high-velocity air jets are dispersed in the hot gases and are subjected to flash combustion. The bottom of the combustion chamber consists of a moving grate that collects inert materials and heavy combustibles after they are blown into the boiler. These materials settle on the end of the grate away from the feed end and gradually move towards the feed end. The grate speed is variable but generally a residence time of about one hour is provided. An upward airflow through the grate provides for completing combustion, during the first two thirds of the travel, and for partial cooling of the ash towards the end of the travel. This counter-current heat exchange mode allows good heat recovery and eliminates the need for quenching the bottom ash with water. The temperatures reached in the middle of moving bed are high and the discharged "bottom" ash, collected by one of the co-authors and examined in the laboratory, is semi-fused, unlike the powdery ash of mass-burn plants. During visits of the SEMASS plant, it was observed that the temperatures within the combustion chamber were much higher than for mass-burn plants. It was not possible to look into the combustion chamber with the naked eye, much as is the case in copper smelting and other high temperature reactors.

2.3 Current Developments

In the last years the flue-gas cleaning has been improved and state-of-the-art technology guarantees meeting emission limits. The current priority is the optimization of the thermal process to:

- increase the energy efficiency,
- reduce the flue gas flow
- minimize the development of hazardous substances like dioxins, CO and NO_x
- avoid or minimize corrosion.
- improve the ash management.

Current developments to reach these objectives are:

- Water-cooled grate elements.
- Flue gas recirculation.
- Oxygen enrichment of the primary air.
- Cladding of the boiler tubes.

2.4. Thermal Treatment of Hazardous Waste for Heavy-Metal Recovery

The process is characterized by a separation of heavy metals and residues during the thermal treatment under a sub-stoichiometric atmosphere in a rotary kiln. After leaving the rotary kiln the separated heavy metals are precipitated in a hot-gas ceramic filter. Using this technology, hazardous materials, both liquids and pasty hazardous waste containing heavy metals, can be treated and a product in a quasi raw-material condition can be formed. The process for the recovery of heavy metal oxides and oxide mixtures is mainly appropriate for residues, sewage and sludge from the following industries:

- surface refinement,
- chemicals,
- cosmetics,
- automotive.

The plant for heavy metal oxide and oxide mixture recovery mainly consists of the following components: evaporation plant for the liquid input materials, rotary kiln, hot-gas filter, scrubber column and dioxin filter. The structure of such a pilot plant at Industrieofenbau Aue GmbH is shown in Figure 2.4.

This disposal route for sewage contaminated by heavy metals could be a practical option for small businesses in particular, since it obviates the need for very expensive wastewater treatment.

Figure 2.4. Structure of the Industrieofenbau plant.

2.5. Emission Control

MSW incineration plants are often subject of controversy, usually because of their flue-gas emissions. In the 60s only few people cared about the emissions from incinerators into the air. But in the 70s air pollution became a high priority issue and in the 80s the public focused on pollutants like dioxins and furans, heavy metals and HCl. Emission limits came into force and to meet these regulations operators have invested large sums in flue-gas treatment devices like acid-gas scrubbers, electrostatic precipitators or baghouse filters. Table 2.2 shows the limits of the actual Waste Incineration Plant Ordinance:

Table 2.2. Emission limits for waste incineration in Germany

Parameter	Unit	Emission limit
Dust	mg/Nm^3	10
TOC	mg/Nm^3	10
HCl	mg/Nm^3	10
HF	mg/Nm^3	1
SO_2	mg/Nm^3	50
NO_x	mg/Nm^3	200
CO	mg/Nm^3	50
Cd+Tl	mg/Nm^3	0.05
Hg	mg/Nm^3	0.03
Metals	mg/Nm^3	0.05
Dioxins/furans	ng TE/Nm^3	0.1

Figure 2.5 shows the development in flue gas-cleaning until the early 90s. The modern incineration plants use nowadays more simple cleaning devices. Figure 2.6 shows the semi-dry scrubber, a configuration used in modern plants. In order to clean mercury and dioxins

activated carbon is sprayed in as well. To reduce NO_x ammonia is blown into the burn-out chamber in a temperature range of 950°C. The modern type of flue-gas cleaning device has the disadvantage of producing more residue because of the surplus of reagent which is necessary to ensure a perfect performance. Because of the cheap underground disposal mines in Germany (mainly old end-of-life salt mines) the lower investment is overruling the costs of reagent and disposal.

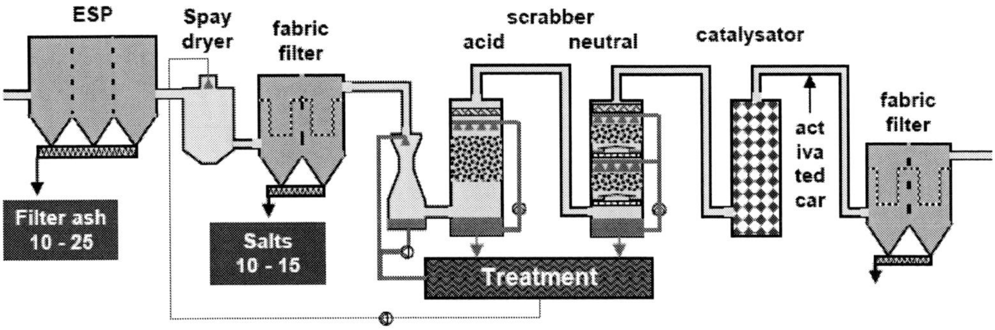

Figure 2.5. Flue gas cleaning system development until 1990 to 1995 (Vehlow, 2006).

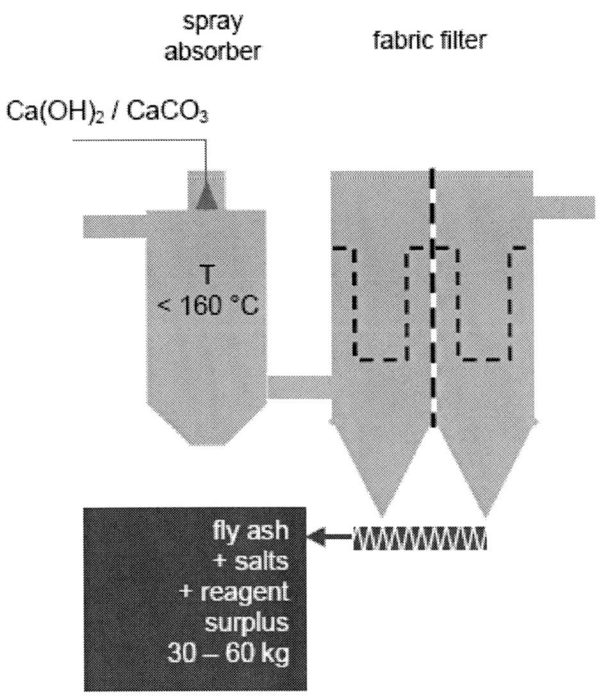

Figure 2.6. Semi dry scrubber (Vehlow, 2006).

Another possibility of a dry cleaning near stoichiometric conditions is the use of natrium hydrocarbonat, which is used in some of the French and Italian incineration plants. The amount of dioxins in Germany dropped from 1990 to 2000 to 1/1000. In 1990 one third of all dioxin emissions in Germany came from waste incinerator plants, whereas for the year 2000 the contribution was less than 1% as shown in Table 2.3.

Depending on the waste composition, MSW consists of a biogenic or renewable part from 57 % to 73 % by weight. The main biogenic fractions are yard and kitchen waste, paper, cardboard and wood. Based on the composition and the biogenic and fossil parts, the $CO_{2,fossil}$ – emission factors (relating to the weight) results for household waste in several European countries are illustrated in Figure 2.7.

Table 2.3. PCDD/F emission sources in Germany (www.umweltbundesamt.de)

sources	Emission per year in g TE		
	1990	1994	2000
metal industry	740	220	40
sintering plants	575	168	< 20
iron- and steel production	35	10	< 5
waste incineration	400	32	< 0,5
municipal solid waste	*399*	*30*	*< 0,4*
hazardous waste		2	0,04
medical waste		0,1	0,0002
sewage sludge		< 0,1	0,03
power plants	5	3	< 3
industrial combustion facilities	20	15	< 10
domestic stoves	20	15	< 10
traffic	10	4	<1
crematoria	4	2	< 2

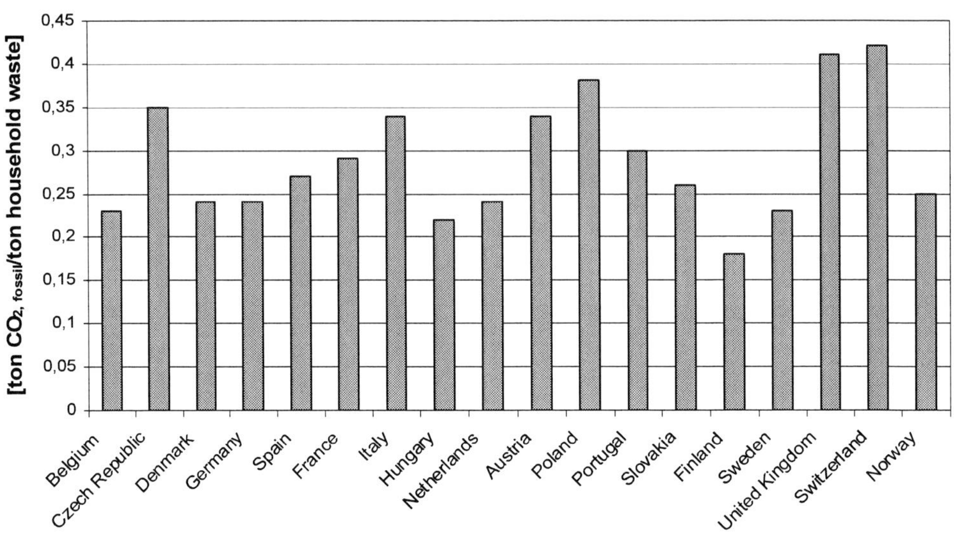

Figure 2.7. Fossil CO2 emission factors (by weight) for MSW in European countries.

With respect to the fossil carbon content and the respective heating values, the emission factors for household, commercial and bulky waste are illustrated in Figure 2.8. The emission factors and heating values of selected fossil fuels are also presented to make a comparison possible. It is obvious that the emission factors (related to the energy content) of the waste types are clearly below those of the fossil fuels but this does not mean that the waste incineration with energy recovery always obtains a CO_2 reduction. The reduction potential by

WTE plants depends on the fuel and the efficiency of the energy production compared to the alternative use as, for example, RDF in power plants.

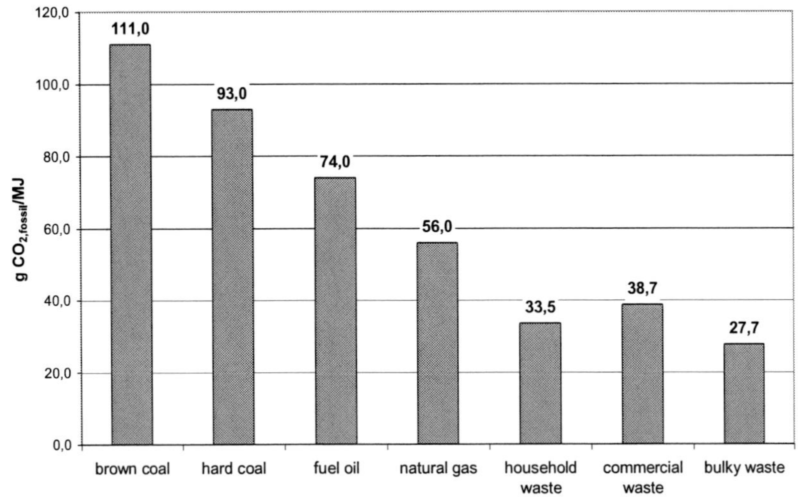

Figure 2.8. Fossil CO_2-emission factors (by energy content) for several waste types and fossil fuels.

The reduction of CO_2 is comparatively cheap compared to the use of wind and solar energy. The following example gives a good overview (Bilitewski et al., 2005):

The increase of efficiency of electric power of 5 % will have CO_2 avoidance costs of 43 €/t CO_2, the increase of heat-use by 8 % will produce CO_2 avoidance costs of approx 7-20 €/t CO_2.

The avoidance costs of alternative energy production are much higher:

- Wind energy: 90-180 €/t CO2.
- Solar energy: 1800-2300 €/t CO2.
- Biomass use can have a price of up to 60 €/t CO2 avoidance.

Section 3: Energetic Recovery of Refuse-Derived Fuel

3.1. RDF Production

Residual waste designated suitable for treatment, such as packaging made of paper, glass, plastics and metal, as well as biological waste, are reprocessed as materials or are re-used for the production of energy. Following the realization of individual collection of re-usable materials and biological waste, current planning is directed towards achieving energetic use of residual waste by producing Refuse Derived Fuel (RDF). The most important characteristics for RDF as a fuel are the heating value, water content, ash content and ash melting point, sulphur and chlorine content. The use of RDF in industrial combustion processes requires quality standards of the mentioned parameters. Figure 3.1 illustrates the production scheme of RDF from MBT plants.

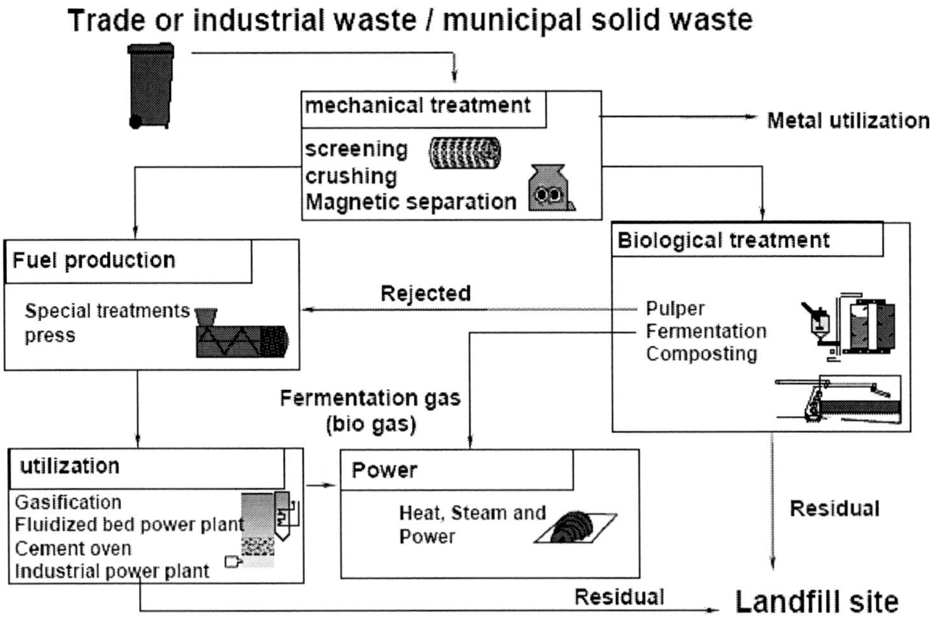

Figure 3.1. Production of RDF from MBT plants for energy use (Keldenich and Marzi, 2006).

3.2. Expected Quantities in Europe

The production of RDF [lately also called SRF (Solid Recovered Fuel)], from household and business waste has grown from 2000 to 2005 from 1.4 million to 12.4 million tons/year as shown in Figure 3.2. Up to the year 2007 the total amount of RDF produced can be estimated to approx. 15 million tons/y in Europe of which Germany alone produced approx. 7 million. The incineration of this amount of RDF is not easy.

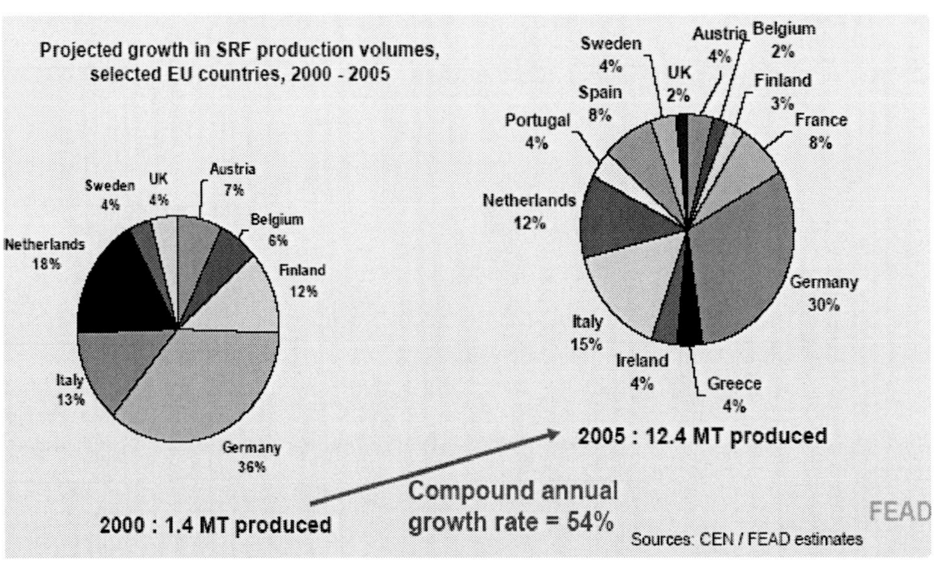

Figure 3.2. Growth of RDF in Europe.

Table 3.1. Survey of quality standards for substitute fuels

	LAGA (proposal) Working Group on Waste by the German Bundesländer Criteria for energetic utilisation in cement kilns	BUWAL Federal Office for Environment, Forest and Landscape Swiss guideline of waste disposal in cement kilns	BGS (proposal) German Quality Association for Secondary Fuels Quality standards for substitute fuels
	(LAGA, 1997)	*(BUWAL, 1998)*	*(Püchel, 1999)*
	mg/MJ	mg/MJ	mg/MJ
As	1,9	0,6	0,5
Be	0,13	0,2	0,13
Cd	0,3	0,1	0,5
Co	1,2	0,8	0,75
Cr	3,7	4,0	15,09
Cu	3,7	4	35,21
Hg	0,02	0,02	0,05
Ni	3,5	4	7,55
Pb	k.A.	8	12,58
Sb	0,07	0,2	3,02
Se	0,2	0,2	0,5
Sn	0,4	0,4	7,55
Te	0,04	k.A.	0,4
Tl	0,15	0,12	0,15
V	6,7	4	1,51
Zn	8	16	k.A.
Chlorine	1 % by weight	n.a.	n.a.

Basis for conversion of mg/kg dry into mg/MJ: Lower heating value (dry) 18.000 kJ/kg.

In Germany roughly one million tons of RDF is kept on stock without the chance to be burned in the next few years. The lack of capacity for the incineration of RDF has a number of reasons with fuel quality being the most important.

3.3. Quality

Up to today, the market for substitute fuels has developed slowly. It is essentially influenced by three participating groups, the producer of substitute fuels, the users of substitute fuels and the approving authorities. The existing dissent of interest between these groups is clearly reflected in the discussion about quality standards for substitute fuels. While the authority represents primarily ecological interests and therefore judges quality standards by the emission situation and the transfer into different environmental compartments, the producers of substitute fuels are in the role of waste processors. They define quality standards on the basis of waste as a raw material and its associated pollutant load. For the user, rather technical aspects are in the centre of interest, especially the operational reliability and aspects of product quality. These contrary interests are also reflected in the ongoing discussion on the assurance of quality of the production and use of substitute fuels. From different sides, standards for substitute fuels are presently suggested.

The Federal Office for Environment, Forest and Landscape (BUWAL) of Switzerland regulates the use of wastes in cement kilns by a guideline which gives advanced specifications. Suitable waste sorts are defined in a positive list. Wastes that are not mentioned in the list must not exceed certain energy-referred pollutant limits.

Waste from households and subsequently segregated fractions are principally excluded. Basis of these regulations are considerations of material flows, which follow two substantial guidelines:

- no increase of the entire emission load from the production of cement,
- no enrichment of the pollutants in the product (clinker),

During the late 70s and early 80s the idea to produce fuels from household waste failed not at least because of unsolved pollutant problems in co-combustion processes. Due to high chlorine and heavy metal content the fuel products could not create sufficient demand. Before the decision is made to process high caloric fractions to marketable fuels in the future, it must be discussed whether the load of pollutants in products and goods has decreased so far, thus leading to a pollutant reduction in the waste stream. For an evaluation, the limit guide values published by BUWAL can be used (Figure 3.3).

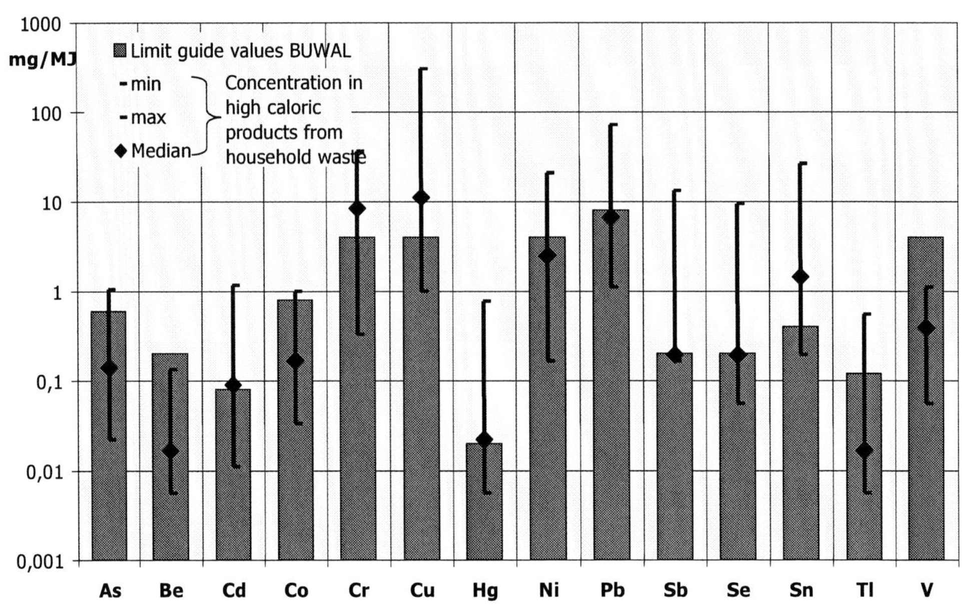

*Basis for conversion mg/kg dry into mg/MJ: Lower heating value (dry) 18.000 kJ/kg.

Figure 3.3. Heavy metal concentration of high caloric fractions from today's mechanical-biological waste treatment plants (literature evaluation*), (Ketelsen, 1999)) in comparison with the BUWAL limit guide concentrations considering the insert into cement kilns (BUWAL 1998).

All maximum values exceed the limit guide concentrations except for exotic elements such as beryllium or vanadium. Particularly relevant are however those heavy metals, whose median is situated clearly above the concentration limits. From this point of view, cadmium, chrome, copper, mercury, lead, antimony and tin can be regarded as guide parameters for the potential of pollutants in MSW. In some cases, chlorine is a limiting factor for the quality assurance of substitute fuels too. The missing limitation of this parameter is justified by the BUWAL by the fact that the chlorine concentration is limited not for ecological reasons, but primarily for technical reasons by the plant operators on < 1% by weight. Due to chlorine concentrations of 0,5-3 % (dry basis) (Ketelsen, 1999), high caloric fractions cannot keep this limiting value in all cases.

Table 3.2 lists theoretical characteristics of RDF. The calculations are based on screening (screen only) as well as screening and separation of heavy materials (screen+heavy). Screening sizes are 40 mm (medium) and 100 mm (coarse). Table 3.3 compares several refuse-derived fuels with the characteristics of primary energy types and the demands of the cement industry. A comparison between Table 3.2 and Table 3.3 highlights the fact that a further reduction in the level of harmful substances is feasible. Practical tests have confirmed these theoretical calculations.

Table 3.2. Theoretical calculation of levels of harmful substances in RDF

(All values for dry matter)	Theor. RDF (medium) (screen only)	Theor. RDF(coarse) (screen only)	Theor. RDF(medium) screen+heavy	Theor. RDF(coarse) screen+heavy
Chlorine overall in %	0.89	1.89	0.77	0.59
S overall in %	0.26	0.33	0.3	0.28
Pb in mg/kg	1600	3770	130	170
Cd in mg/kg	3.3	7.5	1.8	1.0
Cr in mg/kg	290	340	66	86
Hg in mg/kg	2.55	3.87	0.23	0.14

Table 3.3. Characteristics of primary and secondary fuels

(All values for dry matter)	Brown coal literature	Brown coal VEAG	RDF in the cement industry	RDF Velsen	RDF Herne	Dry stabilized residues
				(Bünsow and Dobberstein, 1985)		(Heering et al., 1999)
Calorific value kJ/kg	8,000-20,000	8,496-8,854	13,000	14,560 –14,820	13,130	16,000
Chlorine overall in %	< 0.06	0.01-0.02	<1.5	0.76-0.97	0.56	0.44
S overall in %	0.3-5	0.56-0.7	<0.4	0.15-0.19	0.14	0.45
Pb in mg/kg	7.9-34.1	1.3-7.1	<150	212.4-401.8	138.1	230
Cd in mg/kg	0.1-0.23	0.2-4.5	<10	5.32-6.12	4.45	2.2
Cr in mg/kg	28.1-91.1	6.12-13.5	<150	90.5-127.8	67.1	60
Hg in mg/kg	0.07-0.19	0.1-0.54	<1.5	0.84-1.08	1.25	0.75

3.4. Utilization

3.4.1. Co-Combustion in Power Plants

Power plants based on hard coal or lignite use their fuel in powdered form, briquettes or in form of small milled grain. The types of furnace are optimized in respect of the used type of coal or lignite. These circumstances determine the quality of RDF used in different kind of power plant. In Table 3.4 the specifications of RDF of different power plants in Germany are listed.

The power plants have the biggest potential for co-incineration of RDF if one calculates a substitution of 5 to 10 % for coal and lignite in Germany alone; one could burn between 6.9 up to 13.7 million tons/y. More than 40 power plants have the authorization to use 4.65 million tons of RDF per year in Germany. It is important to notice that burnt in these power plants already are: 490,000 tons/y of waste from the chemical industry, approx. 420,000 tons/y of wastewater sludge, 150,000 tons/y of sludge from paper mills and approx. 420,000 tons/y of other wastes. In 2006, the total amount of RDF from household waste which has been authorized for co-combustion amounted to 2.1 million tons/y. But in reality, most power plants did not burn the authorized amount of secondary fuel from MBT and sorting plants, as only 38% of the authorized capacity was used. The main reasons for power plants not to use RDF from MSW were:

1. Unknown technical risks in the long run like corrosion, higher maintenance, availability etc.
2. Unexpected bad quality of RDF (particle size too big, chlorine and ash content too high, melting point too low, etc.).
3. Market value of gypsum from the flue-gas cleaning system changed dramatically to the worse, so that is could not be sold.
4. Investments in the power plants for handling, shredding and storage were not covered by the tipping fee.
5. Acceptance of the local inhabitants was lacking.

To cover the cost and the risks, power plants in Germany expect tipping fees in the range 48-65 €/t in 2006/7.

Table 3.4. RDF specification of of different power plants in Germany (Eckardt, 2005)

Parameter	Hard coal	Lignite
water content	10-20 %	20-25 %
ash content	10-25 %	5-40 %
Sulphur	0.5-1.0 %	0.4 -2.0 %
Chlorine	0.4-1.0 %	0.4-1.0 %
Fluorine	0.05-0.2 %	0.05-0.2 %
Cadmium	0.1-10 mg/kg	
Mercury	1.0 mg/kg	
Chromium	150 mg/kg	

3.4.2. Cement Kilns

As 31 out of 49 cement kilns in Germany have the permit to burn 2,673 million tons/y of secondary fuel, 22 % of the energy demand of the cement producing industry is covered by secondary fuel, with commonly used wastes including waste tyres, waste oil, waste wood, plastic and textile wastes. Less than 10 % of RDF from MBT and sorting plants is used as energy substitute. Quality requirements for RDF used in cement kilns are shown in Table 3.5.

Table 3.5. Specification of RDF incinerated in cement kilns in Germany (Eckardt, 2005)

Parameters	Primary incinerator	calcinator
water content	10-30 %	20-25 %
ash content	10-25 %	10-20 %
sulphur	0,7-1,0 %	n.a.
chlorine	0,4-1,0 %	n.a.
fluorine	0,05-1,0 %	n.a.
chrome	120 mg/kg	
antimony	60 mg/kg	
vanadium	25 mg/kg	
cadmium	9 mg/kg	
beryllium	2,0 mg/kg	
mercury	1,2 mg/kg	

The main reasons that the cement industry has been avoiding using RDF from MSW are:

1. Quality specification are not met (chlorine content too high, heating value too low, metal content too high).
2. The additional investment for storage, handling and chlorine by-pass is too high.
3. The RDF ashes do not match the recipe for high quality cement

To cover costs and risks, the cement industry is expecting tipping fees in the range of 30-50 €/t in 2006/7.

3.4.3. Dedicated Incineration Plants

Because of the difficulties of acceptance of RDF in power plants and in cement kilns, a number of specially designed plants for RDF have been built in Germany. 530,000 tons/y RDF are currently burnt in fluidized bed reactors (60%) and grate firing plants (40%). Most of these incineration plants are connected with industrial plants with a high demand of steam and electricity on a 24h-basis (e.g. paper mills). An additional capacity of 2.72 million tons/y is currently in planning or under construction, with the fluidized bed technology covering 1.49 million tons/y out of the total of 2.72 million tons/y. As proven technologies both grate and fluidized bed are able to incinerate RDF. The water-cooled grate is the right choice for this material, whereas both the Circulating fluidized bed (CFB) as well as the stationery bed is used for RDF. The main restriction of fuel from waste to be incinerated in the fluidized bed is the particle size. To ensure that the incineration process is working without problems, the particle size should not exceed 100 mm and no metal or heavy material like stone, glass etc. should be present in the fuel. On the grate, particle size can be bigger and the restriction on the feed material is normally less strict. The fluidized bed has a higher efficiency and fewer problems with NO_x and volatile heavy metal, but often more problems with erosions. Both plant types have problems with high temperature corrosion depending on the amount of chlorine in the RDF. The classic solutions are cladding of the boiler tubes, flue gas recirculation and/or reduction of steam parameters.

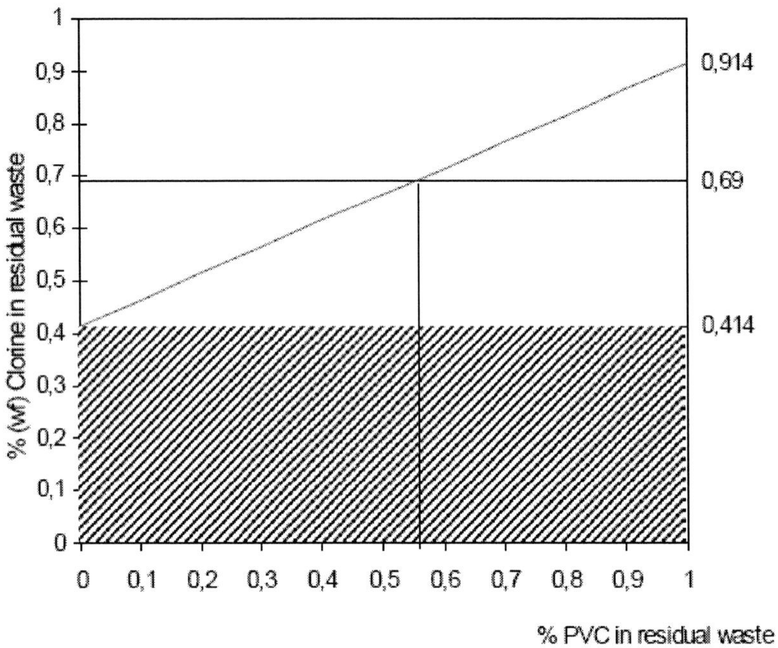

Figure 3.4. Average inorganic chlorine content of 0.414 % as a basic chlorine content in residual waste in Germany and the total amount of chlorine in the residual waste in respect of the PVC content in the waste, resulting in an average chlorine content for the residual waste of 0.69 %.

The new measures are the pre-sorting of the RDF with NIR to identify PVC or other chlorine containing material (PCP, PCB and Chlorphenol). Very often these materials, mainly plastics, contain also heavy metals in over-proportional amounts. Nevertheless, chlorine is still a serious problem when RDF is incinerated, because the corrosion is not so much connected with PVC as it is often expected. The main corrosion is connected with the proportion of inorganic chlorine compounds like KCl, NaCl and $CaCl_2$. This inorganic chlorine content of 0.414 % which is the basis line in Figure 3.4 is not detectable with NIR.

To reach the average chlorine content in the residual waste of 0.69% in Germany (Schirmer, 2007), which did not change over the last 15 years, 0.55% of PVC is approximately needed. But to describe the corrosion properly, the sulphur, calcium, silicium and aluminium content in the RDF must also be known, since all of them have a corrosion inhibitive effect in the incineration process (Schirmer, 2007).

Especially designed RDF incinerators have less problems with chlorine and heavy metal. This is the reason why in Germany a number of specialized incinerators for RDF are under construction. New developments are the gasification of RDF combined with a power plant or cement kiln like those shown in Figure 3.5. The main problem with this combined plant is the availability of the gasification units.

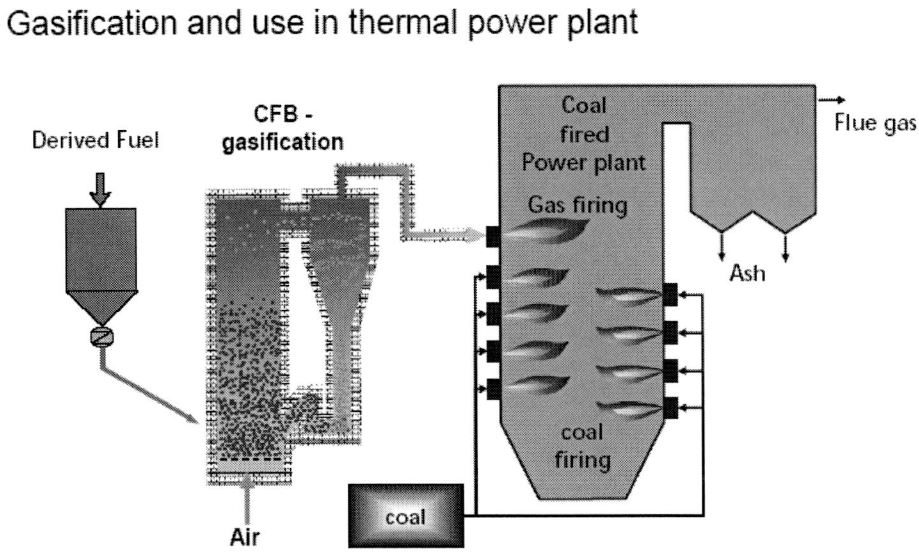

Figure 3.5. Combined coal-fired power and RDF-gasification plant (Keldenich and Marzi, 2006)

3.5. Economic Benefits

For the potential RDF-user, the partial substitution of fossil fuels is interesting only under the condition of an improved proceeds situation. Co-burning of RDF requires first additional investments (handling, storage, fuel supply, if necessary flue-gas cleaning, etc) that are to be transacted by the plant operator. In addition, efficiency losses and increased maintenance cost are possibly to be considered during the usage. These costs are to be represented by a saving of fuel costs and by a possible trading of emissions. Therefore, the following parameters considerably determine the economic efficiency of an RDF employment:

- Cost of the primary fuel.
- Amount of additional payment for RDF paid to the utilization plant by the RDF-producer.
- Cost of the emission certificates.

For the co-combustion of RDF in power plants the tipping fee amounts in 2003 to 7-30 €/t and increased until now to 35 and 65 €/t.

In Figure 3.6, the influence of these parameters is demonstrated exemplarily for the substitution of lignite. To simplify, the calculation is based on a RDF with a average heating value of 16 MJ/kg and an emission factor of 30 g CO_2/MJ. These values will vary for each preparation process due to different RDF compositions. With an assumed fuel price of 2 €/GJ and an RDF additional payment of 10 €/ton, paid to the utilisation plant by the RDF-producer, the saving results to 2.63 €/GJ_{subst}. In the case of an additional possible emission trade with a certificate price of 5 €/ton $CO_{2,}$ this saving increases with lignite to 3.03 €/$GJ_{subst.}$.

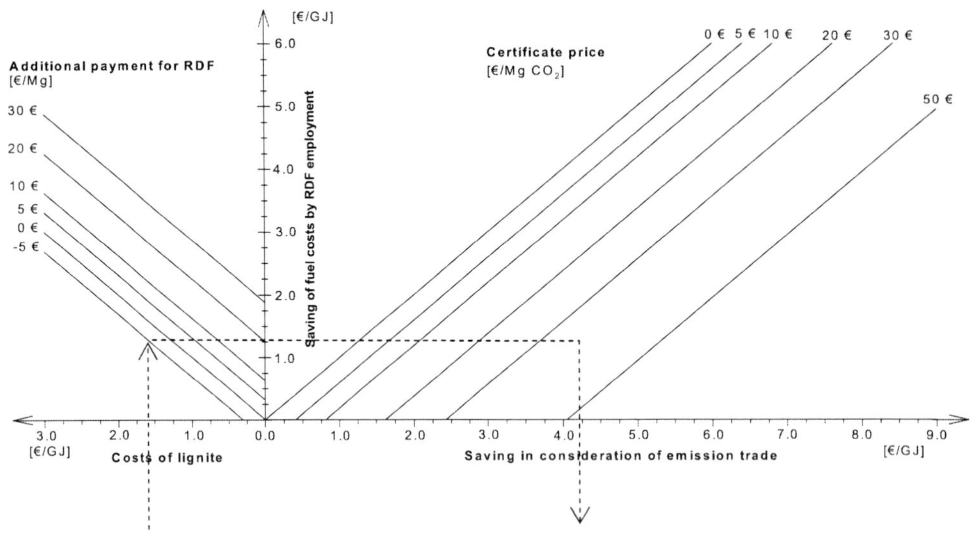

Figure 3.6. Economic effects by the substitution of lignite through RDF.

3.6. Reduction Potential of Greenhouse Gases

Emission factors are an indicator for the average emission of a pollutant related to a defined energy quantity. There are standard values for fossil fuels. For RDF, the emission factors have to be calculated. With respect to the fossil carbon content and the respective RDF heating values, the following emission factors result:

- Biological stabilisation: 18.9 g $CO_{2,fossil}$/MJ
- Mechanical Pre-treatment: 31.6 g $CO_{2,fossil}$/MJ

For comparison, the emission factors and heating values of selected fossil fuels are represented in Table 3.6. It becomes obvious that the emission factors of the regarded RDF are clearly under those of the fossil fuels. In the case of an increase of the plastic portion in the RDF, this is however due to an increase of the emission factor for this RDF.

Table 3.6. Emission factors of fossil and regenerative fuels

Fuel	Emission Factor [g CO_2/MJ]	Lower Heating Value
lignite[1]	111	8.6 GJ/Mg
mineral coal[1]	93	29.7 GJ/Mg
fuel oil[1]	74	35.4 GJ/m^3
natural gas[1]	56	31.7 MJ/m^3
RDF processed after a biological stabilisation	18.9	15.5 MJ/kg
RDF processed by mechanical pre-treatment	31.6	17.4 MJ/kg

[1] according to UBA, 1996.

Figure 3.7 illustrates the saving of fossil fuels and the reduction potential of CO_2 by the energetic use of the calculated RDF quantity. It is assumed that the entire RDF-quantity is used in each case for the substitution of a fossil fuel. It is shown that, for reasons of climate protection, the substitution of the energy source with the highest emission factor (lignite) appears most meaningful. In practice, however essentially economic reasons will determine the RDF use.

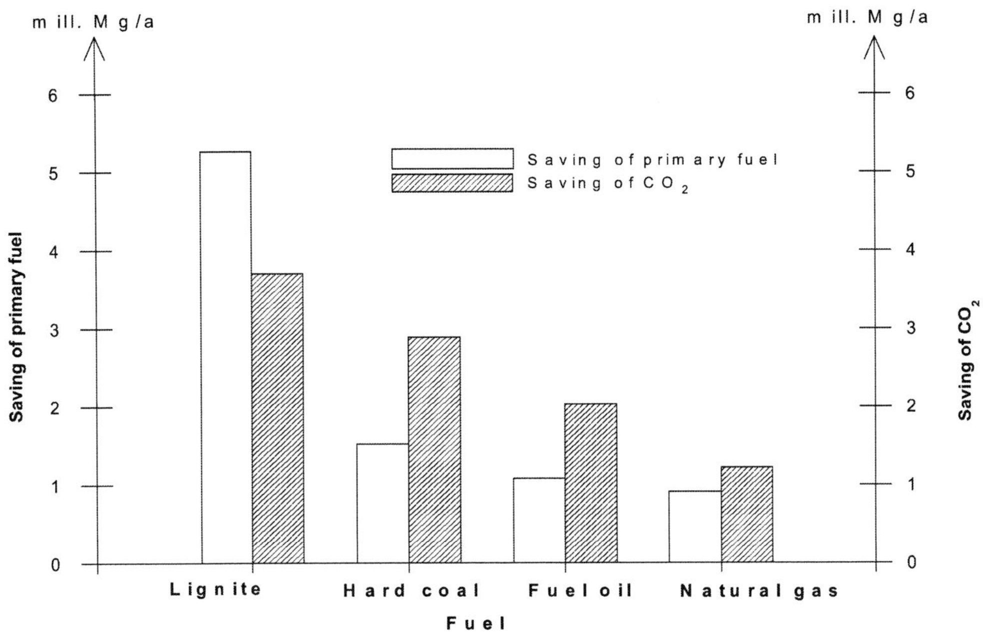

Figure 3.7. Saving of fossil fuels and the reduction potential of CO_2 by the use of RDF.

SECTION 4: PYROLYSIS AND GASIFICATION

4.1. Industrial Roots

Pyrolysis is a well known process from industrial process engineering. The further development of pyrolysis and its application to waste was prompted by the drawbacks experienced with MSW incineration. With respect to the treatment of MSW, pyrolysis was expected to provide the following benefits (Bilitewski et. al., 1997):

- Uncomplicated and cost effective processes that operate with feed rates as low as 10 t/h.
- A potential for energy and resource recovery.
- The recoverable product (energy) is easily stored.
- Flexibility with respect to various wastes and changing waste compositions.
- Almost complete prevention of environmental pollution.

Waste gasification and pyrolysis technologies are used to pursue similar benefits as are derived from waste incineration. Waste materials are significantly reduced in volume and are converted to a physical state that permits their storage without serious effects on the environment. The relationship between various thermal processes such as drying, pyrolysis, gasification, and incineration are presented in Table 4.1.

Table 4.1. Classification of various thermal treatment methods (Rasch, 1977)

Potential process steps	Treatment method			
	Drying	Pyrolysis	Gasification	Combustion
Step 1	Drying	Drying	Drying	Drying
Step 2		Pyrolysis	Pyrolysis	Pyrolysis
Step 3			Gasification	Gasification
Step 4				Combustion

During conventional incineration, the processes of drying, pyrolysis, gasification and combustion (burning of gaseous products), take place in a combustion chamber. During pyrolysis, some of these incomplete processes are allowed to take place in separate reactors, which makes pyrolysis and gasification separate waste treatment processes. However, pyrolysis was unable to compete with incineration because of many problems and its insufficient availability. There is only one MSW pyrolysis facility in operation in Burgau, Bavaria, since 1984 and no additional pyrolysis facilities are planned. However, pyrolysis is to be used in combination with high-temperature incineration. Here, the extracted pyrolysis gases are used in a second stage to incinerate or vitrify the pyrolysis coke. The best-known example is the low temperature combustion process.

4.2. Pyrolysis

Pyrolysis represents the thermal decomposition of organic materials by excluding gasification aids such as oxygen, air, CO_2, steam, etc. In the temperature range between 150 and 900°C, volatile compounds are expelled, and complex carbohydrates are broken down into simpler ones. This process is also called low temperature carbonization or destructive distillation. Table 4.2 lists a number of chemical reactions that occur during pyrolysis. Depending on the operating parameters of the facility, the generated pyrolysis products include gas, coke, oil, and tar. The recovery of oil fractions by distillation is economical only in large quantities or with high-grade input such as plastics, discarded tires, etc. Therefore, the main focus of pyrolysis is the production of gas. Depending on the input material, pyrolysis gas yields between 12,500 and 46,000 kJ/Nm^3; however, it is not free of condensable components. The solid residue consists of pyrolysis coke which contains varying amounts of residual carbon that is not converted to gas in the process. This effect reduces the efficiency of the pyrolysis process compared to gasification. Furthermore, pyrolysis is an endothermal process, i.e. energy must be supplied.

Table 4.2. Pyrolytic decomposition of organic materials as a function of surface temperature (Bilitewski et.al., 1997)

Temperature range °C	Chemical Reaction
100 to 120	Thermal drying, dehydration
250	Deoxidation, desulfurization; molecular splitting of water and carbon dioxide; begin cleaving of hydrogen sulphide
340	Breakage of bonds of aliphatic compounds, begin splitting of methane and other aliphatics
380	Carbonization phase, concentration of carbon onto residues
400	Breakage of bonds of carbon-oxygen, carbon-nitrogen
400 to 600	Decomposition of bituminous materials into low temperature carbonization oil and tar
600	Cracking of bituminous materials into heat resistant materials (gaseous, shortchained hydrocarbons), formations of aromatics, (benzene derivatives)
> 600	Olefin (ethylene) dimerization to α-butylene@ dehydrogenation to butadiene, reaction of ethylene to cyclohexane; thermal aromatization to benzene and higher volatility aromatics

4.2.1. Input Materials and Output Products

The pyrolytic conversion of wastes results in a variety of products, their composition being determined by the input material, the heating parameters, the pyrolysis temperature, and the reaction time. Figure 4.1 displays a material and energy balance of 1 kg of MSW burned during low-temperature pyrolysis. The pyrolytic conversion of waste results in a number of different products, the composition of which are determined by the composition of the input material, the conditions during heating to the degasification temperature and by the duration of reaction. Specifically, the following end products are possible:

- Fuels, i.e., raw material in the form of tars, oils, or combustible gases.
- Condensate with dissolved contaminants.
- Residues such as waste coke, metals, glass, sand, etc.

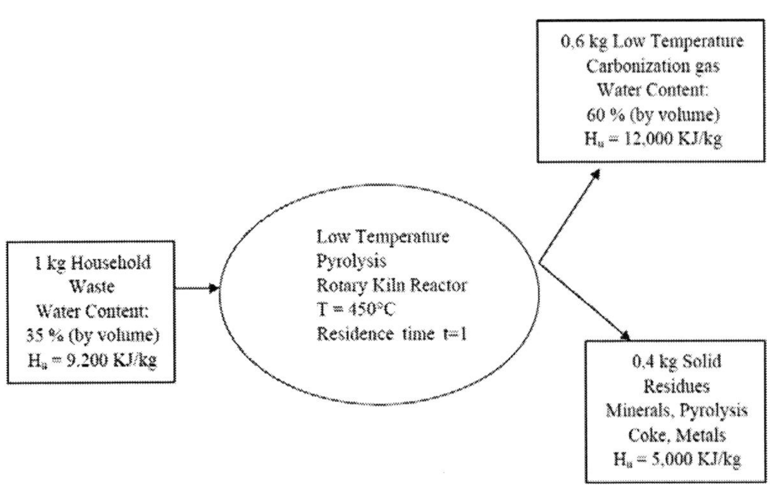

Figure 4.1. Example of material and energy balance from low-temperature pyrolysis of MSW (Thomanetz, 1983).

4.2.2. Marketability of Outputs

There is a limited market for some of the pyrolysis products. In particular, the oils extracted during scrap tire pyrolysis are usable as raw material in chemical or petroleum feedstock. In this case, a necessary requirement is therefore the physical proximity of the pyrolysis facility to an oil refinery; the same applies to the pyrolysis gases, which must be used in part to fuel the facility's own heating processes. Pyrolysis coke, despite a number of potential uses like carbon black or activated carbon has not found any stable or reliable markets (Bilitewski et.al., 1997).

4.3. Gasification

Gasification refers to the conversion of carbon-containing materials at high temperatures into gaseous fuels. Gasification is differentiated from pyrolysis by the addition of reactive gases, which further convert carbonized residues into additional gaseous products. Gasification is, strictly speaking, the continuation of the pyrolysis process, where the residual carbon is oxidized from the glowing embers of the pyrolysis coke at temperatures above 800°C with sub-stoichiometric oxygen. Steam, carbon dioxide, oxygen, or air are used as gasification agents. Just as pyrolysis, gasification is an independent process, but is still also a part of the combustion process. The products generated in the gasification process are determined by the type of agent used, e.g., lean gas, water gas, etc. The necessary reaction energy for the gasification process is generated by the partial combustion of the organic material inside the reactor. Thermoselect and Schwarze Pumpe are both gasification processes.

4.3.1. Input Materials and Output Products

The products generated during gasification are lean gas, water gas, and producer gas. Their quantity and composition are determined by the type of gasification agent as well as the temperature. The high-temperature gasification process, with temperatures ranging between 800 to 1,100 °C provides the highest quantitative yield of gas, but it is gas with a comparatively low heat value. The immediate use of these gases in a secondary combustion chamber is desirable, since this is how the existing heat can also be recovered. Producer gas has a lower heat than pyrolysis gas and in comparison to the waste volume this results in a higher gas volume. After cracking, virtually no condensable constituents remain and the gas can be readily burned since the pure gas contains only a minimum of harmful compounds (SO_2 = 20 mg/Nm^3, HCl = 23 mg/Nm^3 and HF = 1.3 mg/Nm^3. Besides the non-combustible components and depending on the operating temperatures, ash or slag residues are generated. The solid residues from the gasification process are similar to those from incineration; they also have a high percentage of ash and a low percentage of carbon.

4.3.2. Marketability of Outputs

Various tests with natural gas engines, lean gas burners, etc., have been conducted to develop beneficial uses for low-caloric gas from gasification. The immediate combustion in a secondary combustion chamber has proven to be the most efficient solution. In contrast to pyrolysis, gasification slag is vitrious and does not leach contaminants into the environment.

Therefore, this material is suitable for use in trail and road construction, as well as for winter road service (as sanding material).

4.4. Environmental Considerations

With respect to environmental effects from pyrolysis processes, a distinction can be made between contaminants inherent to the medium and those specific to the process. Medium-inherent contaminants results from elements such as sulphur, fluoride, chlorine, heavy metals, and mercury, as well as from compounds of these elements. Process-specific contaminants are characterized by the fact that their formation is a result of the thermal process. The type of facility and the operating conditions (temperature, duration, etc.) determine the type of contaminants.

Pyrolysis wastewater consists of moisture from waste, decomposition water and water resulting from combustion, minus the amount that is used in the water-gas reaction. The wastewater leaves the reactor as vapour and then condensates as the gas cools. Also included in the wastewater, besides the condensate, is the liquid from scrubbing and the cooling water from the wet removal of solid residues. The total wastewater quantity is between 0.4 and 2 m^3/ton of the original input.

The contaminated wastewater is high in organics. The pyrolysis of MSW with high concentrations of oils, tars, phenols, etc. in particular, requires chemical/physical pre-treatment, because the contaminants can only be partially decomposed in biological sewage treatment plants. Another possibility of reducing wastewater contamination is the thermal destruction of the pyrolysis gas as described above. The water from wet removal of solid residues can extract soluble compounds (water-soluble calcium, chlorides, sulfates and organic compounds) up to a limited degree.

Pyrolysis of organic materials generates several hundred different polycyclic aromatic hydrocarbons (PAHs). It has been demonstrated that the PAH concentrations in MSW are much higher than in the solid residues of pyrolysis, i.e., pyrolysis residues PAHs in the solid phase. Pyrolysis generates only small quantities of PCDD and PCDF as compared to incineration, because oxygen is necessary for dioxins to form, whereas oxygen enters the pyrolysis process only as a part of the waste.

4.5. Hydrogenation and Hydrolysis

Hydrogenation refers to the reaction of certain substances, e.g., coal and macro-molecular chemicals, with hydrogen under pressure and rising temperatures. Hydrolysis, on the other hand, is a process in which water reacts with a chemical compound to form two or more new substances.

The potential of treating carbon-containing waste with high pressure hydrogenation has been discussed since the 1950s. High pressure hydrogenation was developed in the late 1920s, with the intent of making gasoline from coal. In 1981, a coal-oil hydrogenation plant was built in Bottrop, Northrhine-Westphalia. This plant uses the modified Bergius-Pier-Process (VEBA Combi-Cracking [VSS]) to liquefy hard coal. Aside from coal, other materials can be used, including: distillation residues from crude oil processing, waste oil, waste oil containing PCBs, spent activated carbon (containing solvent), spent paint thinner, and scrap plastics.

Hydrogenation occurs at temperatures between 440 and 480°C, at a pressure of about 300 bars and in the presence of hydrogen. Macro molecules are split into smaller molecules and are saturated with hydrogen. For example, polyolefines, polyamids, and aliphatic polymethanes are converted to gaseous and liquid hydrocarbons. The hydrogenation products then become raw materials for oil refineries. The heteroatoms in the input material, such as oxygen, nitrogen chlorine, and sulphur are separated and form water, ammonia, hydrochloric acid and hydrogen sulfides. It is noteworthy that mixed plastics that have been collected by the DSD in Germany, have been used successfully in the Bottrop plant.

SECTION 5: BIOCONVERSION OF SOLID WASTES

5.1. Anaerobic Digestion and Fermentation

The anaerobic digestion process, carried out in the absence of oxygen, involves the use of microorganisms for the conversion of biodegradable biomass material into energy, in the form of methane gas, and a stable humus material. Anaerobic digestion can occur under controlled conditions in specially designed vessels (reactors), semi-control conditions such as in a landfill, or under uncontrolled conditions as it does in the environment. It should be noted that anaerobic digestion is differentiated from anaerobic fermentation which is the term usually applied to processes employing anaerobic microbes for the production of fermentation products such as alcohol or lactic acid.

Table 5.1. Weight distribution and biodegradability of the organic fraction of MSW found in the U.S.A

	Percent by wet weight		Biodegradability		
Component	Range	Typical	Low[a]	Medium[b]	High[c]
Paper					
Cardboard	3-10	10		x	
Magazines	2-8	6	x		
Newspaper	4-10	8	x		
Waxed cartoons	5-20	12			x
Other (mixed)	<1-5	3			x
Food waste	6-18	9			x
Yard wastes	5-20	15			x
Wood waste	1-4	1.5	x		

[a] Low-biodegradable materials are classified as having a biodegradability of less than 30%.

[b] Medium-biodegradable materials are classified as having a biodegradability greater than 30 percent but lower than 75%.

[c] High-biodegradable materials are classified as having a biodegradability greater than 75%.

Major sources of waste materials considered as a feedstock for anaerobic digestion process are (1) MSW, (2) agricultural animal waste, (3) crop residues, biomass and energy crops, and (4) wastewater treatment plant sludge. The composition of MSW may vary greatly by season, geographical area, and community socio-economic level. The extrapolation of results from one location to another, therefore, may not be valid and should be done with caution. The biodegradability of these waste materials varies substantially as reported in Table 5.1.

5.2. The Issue of Biodegradability

While each organic waste may contain a constant ultimate biodegradable fraction, practical biodegradability can be quite variable. Factors such as particle size, time and environmental conditions (i.e., temperature, nutrient requirements, etc.) will influence the final outcome of biodegradation. For example, because favorable conditions do not exist in most landfills, the biodegradability estimated from analytical test will usually be greater than the actual biodegradation that occurs in landfills. For practical purposes, the discussion in this section deals with the biodegradability of the organic fraction of waste materials in an in-vessel anaerobic digestion process. Substrate biodegradability is of special importance in an in-vessel anaerobic digestion process, where the production of energy is of concern.

5.3. Biodegradable Volatile Solids of an Organic Substrate

Dry organic substrates consist of volatile solids (VS), and ash. Taken together these two components comprise the total solids (TS) of a substrate. The fraction of ash typically depends on the nature of the organic substrate. VS is measured as loss on ignition. Only the biodegradable volatile solids (BVS) fraction of the VS has the potential for bio-conversion largely because of the presence of refractory volatile solids (RVS) which, in most digester feedstocks, is mostly lignin. Lignin is a complex organic material which is difficult for anaerobic bacteria to degrade and normally requires a long period of time for complete degradation. The RVS in the organic fraction of MSW consists of the lignin content which is associated with cellulose in plant materials and thermo-plastic materials. Lignin content of an unsorted organic fraction of MSW will fluctuate as the percentage of cellulose and thermo-plastic materials will vary with season, socio-economic conditions and geographical locations. Knowing the BVS fraction of VS in the individual organic fractions of MSW will allow the biodegradability of any composite biodegradable organic fraction of municipal solid waste (BOF/MSW) to be estimated.

5.4. Anaerobic Bacteria

Effective anaerobic degradation of complex organic waste is a result of the combined and coordinated metabolic activity of the digester microbial population. This population of microorganisms is comprised of several major trophic groups that process different carbon catabolizing functions. At present, four different groups of anaerobic bacteria are recognizable as reported in Table 5.2. With respect to methane recovery, the anaerobic bacteria are generally grouped as: (1) hydrolysis (hydrolizing bactera), (2) acetogenes (acid forming bacteria), and (3) methanoges (methane forming bacteria).

Table 5.2. Four major groups of anaerobic bacteria and their functions

Bacterial group	Function
Hydroletic Bacteria	Catabolize saccharides, protein, lipids, and other minor chemical constituents of biomass.
Hydrogen Producing Acetogenic Bacteria	Catabolize certain fatty acids and neutral end products.
Homoacetogenic Bacteria	Catabolize unicarbon compounds (e.g., H2/CO2 or HCOOH) or hydrolyze multicarbon compounds to acetic acid.
Methanogenic Bacteria	Catabolize acetate and one carbon compounds to methane.

Methane bacteria are among the most strictly anaerobic microorganisms known and occur naturally in the rumen of cows, in marshes and in brackish waters, as well as in wastewater treatment plant digesters. When the bacteria are isolated in pure culture, some of them have been shown to be able to manufacture all of their cell constituents from simple carbon compounds. Their growth can be enhanced to some extent by the further addition of the keto acid pyrovate which stimulate extra supply of carbon source.

5.5. Characteristics of Biogas

The gas produced during anaerobic digestion of biodegradable organic material in a healthy fermentation system, called biogas, consists mainly of a mixture of methane (CH_4) and carbon dioxide (CO_2) with small amounts of other gases, including hydrogen sulfide (H_2S), hydrogen (H_2), nitrogen (N_2), and low molecular weight hydrocarbons. Typically, digester gas has 50-75% methane and 25-50% carbon dioxide; the remaining gases are present in very small quantities. The composition of biogas, as obtained from various sources, is reported in Table 5.3. As also reported in Table 5.3, a large variation exists in the composition of biogas, primarily due to differences in feedstocks and operating conditions.

Table 5.3. A typical composition of biogas produced from BOF/MSW

Constituent	Percent by volume	
	Range	Typical[b]
Methane, CH_4	50-75	53.0
Carbon dioxide, CO_2	50-25	45.0
Hydrogen sulfide, H_2S	0.01-1.5	0.02
Hydrogen, H_2	Trace-3.5 [a]	1.7
Nitrogen, N_2	Trace-8 [a]	Trace
Other hydrocarbon	Trace-0.05	Trace

[a]The hydrogen and nitrogen gas reported in this table are more commonly found in landfill gases.
[b] Typical biogas composition from the biodegradable organic fraction of MSW.

Because biogas normally consists of a mixture of gases, biogas characteristics must be evaluated for each individual case. However, in many cases the physical characteristics of the three main gas constituents; namely, methane, carbon dioxide, and hydrogen sulfide can be used to characterize biogas.

5.6. Utilization of Biogas

To understand the potential use of biogas, it is important to gain a perspective on its energy potential by comparing it with more familiar uses of energy. The following are some examples of the use of one cubic meter of biogas, at 60 to 70 percent methane content, for common energy consuming purposes (Table 5.4).

Table 5.4. Equivalent uses of biogas

Use	Equivalent use
Cooking	Can cook 3 meals for a family of 5 to 6.
Lighting	Illumination equaling that of 60-100 watt bulb for 6 hours.
Petroleum	Equivalent to 0.76 kg of petroleum.
Car fuel	Can drive a 3-ton truck for 2.8 km.
Motor power	Can run a 1HP-power motor for 2 hours.
Electricity	Can generate 1.25 kW of electricity.

Table 5.5. Uses for biogas produced from the biodegradation of the organic fraction of MSW

Applications	Comment
Fuel in an IC engines	The most common application of biogas, usually used to generate electricity. Modified engines available from Caterpillar, Cooper-Superior, and Waukesha. Modifications include corrosion resistance and proprietary designing. Specialized lubricating oils are necessary, and must be changed more often. Power ratings are usually 5 to 15 percent below natural gas ratings.
Fuel in a gas turbine	Common application of biogas, used to produceelectricity. Modifications similar to those for IC engines are necessary. Biogas gas turbines are made by the Solar turbine division of Caterpillar. Power rating is 10 to 15 percent lower than for natural gas. Turbines must be checked often for corrosion or deposition. Water vapor should be removed from biogas before use.
Boiler fuel	In moderate use, currently, requires little modification to current equipment. A mixture of biogas and natural gas may be used. Boilers appear to be less sensitive to gas contaminants than IC engines or gas turbines.
Pipeline quality gas	Less common use. CO_2. and all contaminant gases, including water vapor, must be removed before gas will be accepted into a pipeline. Usually not economical unless a large amount of gas is produced and a long-term contract for the sale is available.
Fuel in a steam turbine	Currently in limited use. A large amount of gas is necessary before the system will be economical for the production of electricity. However, modifications are small.

Table 5.5. Continued

Applications	Comment
Fuel for space heating	Currently in limited use. Only minor modifications necessary to convert equipment which uses natural gas. Economical even on a small scale. Hampered only by unavailability of gas. Pipelines are not currently economical so only used at production sites.
Fuel for industrial heating	Currently in limited use. Can be used for lumber drying, kiln operations and cement manufacturing. May be used in place of or with natural gas, using natural gas equipment.
Fuel for fuel cell	Technology under development. These open fuel batteries can be designed to use biogas, or biogas can be converted to hydrogen in a catalytic pretreatment unit.
Compressed vehicle fuel	Technology under development. Gas must be purified to near pipeline quality.
Convert to methanol	Technology under development. Currently too costly.
Synfuel or chemical feedstock	Technology under development. Liquid fuels and acetic acid production is being investigated.

The potential commercial uses of biogas as a source of energy are summarized in Table 6.5. Some applications listed in Table 5.5 are currently practiced and others are in the process of research and development.

5.7. Digestion of Solid Substrates

While with wet digestion the solids content lies between 3 and max. 15 %, with dry digestion the substrate is not pumpable (solids > 30 %). In practice there are now many process variations, with no clear line between wet and dry digestion, and including 2-stage processes which combine the two. Both processes are available in mesophilic and thermophilic variations. In dry digestion, a distinguishment between continuous and batch processes must be made (Figure 5.1).

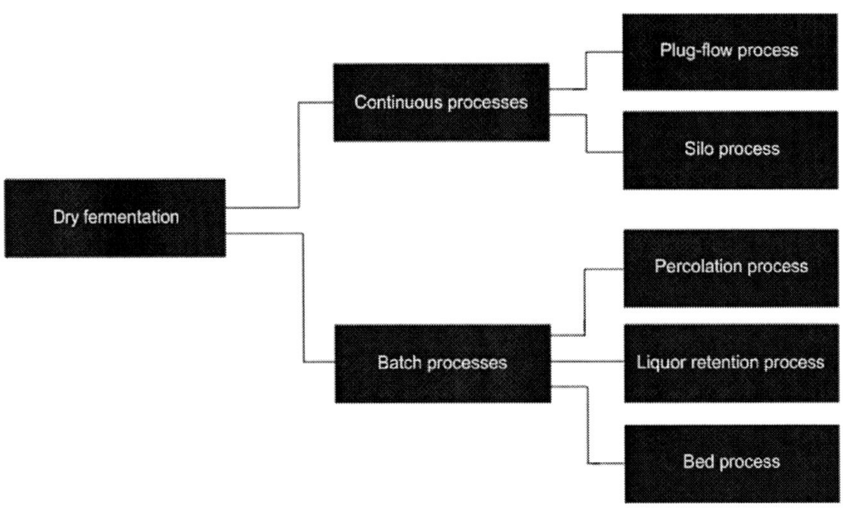

Figure 5.1. Different processes of solids' digestion.

5.8. Examples of Dry and Wet Digestion Plants

Pros and contras for wet and dry digestion plants in the digestion of high-solids substrates will be discussed using the examples of the wet digestion plant, "Radeberg" (conventional concept) and the BEKON dry digestion plant ("Munich", innovative, new process concept). The two processes are briefly described below.

5.8.1. Wet Digestion Plant, Radeberg

In the original design of biogas plants, great value was placed on the flexibility of the plant, the goal being for it to be able to handle as wide a range of substrates (biowaste) as possible. In addition to the usual liquid and semi-solid wastes, such as the contents of fat separators, fed into co-digestion plants, solid biowaste, for example, from municipal separate waste collection (domestic and commercial) also had to be handled. It had to be possible to digest domestic biowaste together with or without sewage sludge. Thus, the plant was equipped with two digesters (Figure 5.2), which could be operated together or separately. The plant was built by Linde KCA of Dresden and went into operation in the summer of 1999. Figure 5.3 shows a basic flow diagram of the co-digestion plant, in which, in the first years of operation, about 40,000 m³ of clarification sludge (solids content 5 – 6 %) and approximately 15,000 tons of biowaste, mainly from municipal collection by the city of Dresden, were processed per annum.

Figure 5.2. Overview of the co-digestion plant, Radeberg.

Clarification sludge, along with liquid, semi-solid and solid biowaste, were delivered to the plant separately. The clarification sludge is delivered in pre-concentrated form. Liquid and semi-solid biowaste (optionally via a rack) can be stored in tanks, some of which are heated. Solid biowaste is delivered into a shallow bunker, where interfering substances are initially removed. Unsuitable material can be separated out here with a wheel loader and disposed of via a different route.

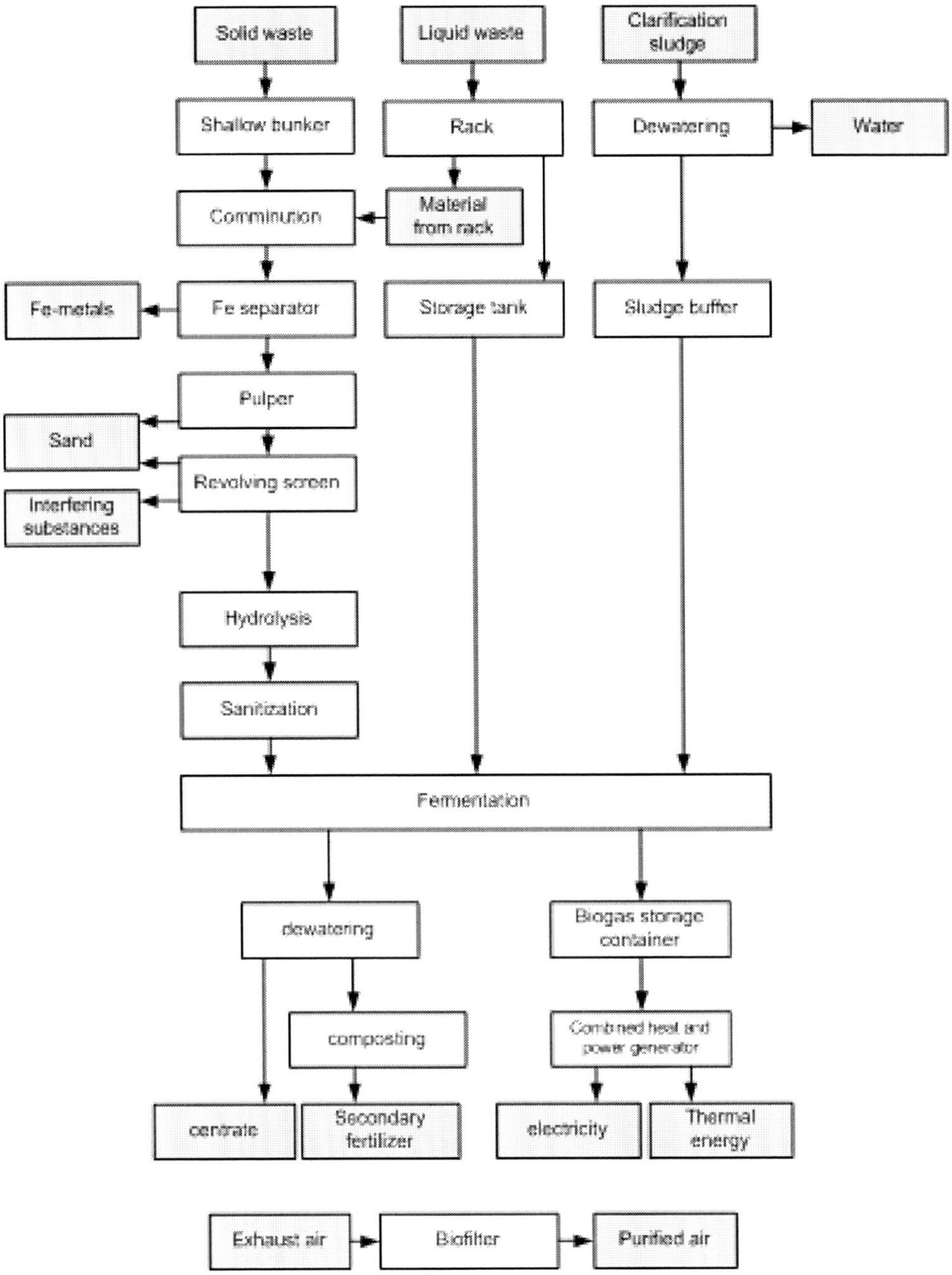

Figure 5.3. Basic flow diagram of co-digestion plant, Radeberg.

From the shallow bunker, biowaste can be deposited by wheel loader into the feed hopper of the comminuting machine. Subsequently, the comminuted substrate is transported via a conveyer belt and through the Fe-removal device to the pulper, where the substrate is further comminuted into pulp. Heavy contaminants (stones etc.) are removed in a gravity trap. Subsequently, the pulped substrate passes over a weir into a revolving screen with segments with differently-sized holes. This is where sand and coarser unwanted material is removed from the suspension, which is then transported to the mash storage tank. From the mash storage tank the substrate in fed via a macerator into a sanitization tank, where it is sanitized, with a grain size of < 10 mm, at 70°C for over 1h. Following sanitization, the substrate is fed into the digester together with clarification sludge. Optionally, the clarification sludge can be digested separately from the biowaste. This is to ensure that by large increases in the price of utilizing clarification sludge, the biowaste can still be sold as compost. Following digestion (mesophile over a period 15-20 days) the residue is dewatered in a centrifuge (the removed water being fed into Radeberg's municipal sewage works) and the solids are also utilized, e.g. in agriculture. The produced biogas is used to generate electricity in a combined-heat-and-power plant. Surplus electricity is fed into the national grid, while the thermal energy is used to heat the reactors (digesters), for the sanitization process, for the sewage works, and for heating plant buildings. Exhaust air from the plant is collected centrally and passed over a bio-filter before being emitted into the atmosphere.

5.8.2. BEKON Dry Digestion

The dry digestion plant of BEKON Technologies GmbH and Co. KG in Munich (Figure 5.4) was originally designed to handle about 8,000 tons of biowaste and is currently being expanded to a capacity of 25,000 tons/y. The plant was constructed in cooperation with the city of Munich, from whose municipal waste collection the substrate comes. A basic flow diagram of the plant is shown in Figure 5.5. The biowaste is deposited on arrival on a sealed but drained surface, where it is stored until taken by wheel loader in batches of about 120 – 130 tons and mixed with roughly equal amounts of already digested biowaste, before being deposited in one of four digesters. Mixing the fresh substrate with substrate that has already been digested serves its inoculation, thus speeding up digestion, which occurs in batch process.

Figure 5.4. Pilot plant for the dry digestion of organic waste.

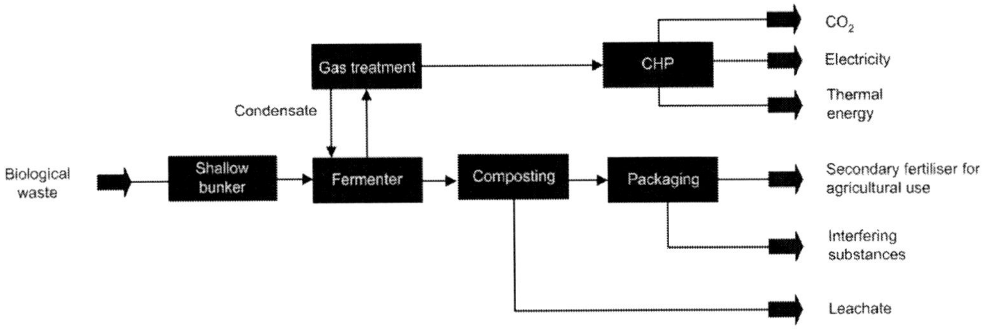

Figure 5.5. Basic flow diagram of BEKON dry digestion pilot plant, Munich.

The biowaste is not given any further treatment prior to digestion. In contrast to many other forms of biowaste treatment (composting, digestion), neither comminution, nor removal of interfering substances, nor any screening takes place.

The digesters are operated in batch process, which means that, after one completed cycle, the digester is completely emptied and filled with fresh (inoculated) material. Once inside the digester, there is no mechanical treatment of the substrate. However, in order to promote the microbial processes involved in the digestion, leachate from the substrate is collected and percolated through it.

Nominally, the retention time of substrate in the digester is about 4 weeks, but because fresh material is mixed in a ratio of 1:1 (other ratios are possible) with already digested material, this increases the mean retention time considerably. If only two digester cycles are taken into account, the retention time is increased from its nominal value by 50 percent (not taking material losses, e.g. biogases, into account), to more than 6 weeks. During this time some of the organic compounds are biologically transformed, larger molecules being reduced, successively, to smaller ones (organic acids, mercaptanes, aldehydes, alcohols, CO_2 etc). Finally, at the end of the anaerobic "food chain", methane bacteria produce highly reduced, energy-rich methane gas (which at up to 65%, is the main component of biogas) from the oxidation of organic C1 and C2 compounds in a process known as carbon respiration.

5.9. Flexibility in the Use of Substrates

The greatest security in respect to substrate supply is obviously provided by a plant that can handle the greatest variety of potential substrates. However, in practice, certain potential substrates can often be disregarded. If, within the catchment area of a plant for processing agricultural substrates there is no source of liquid manure or other liquid biowaste, it is very unlikely that significant amounts of liquid substrate will ever be available for processing. If no liquid substrates are available, the wide range of substrates that can be processed by a wet digestion plant will be of little interest in deciding on the type of plant required. On the other hand, in a dry digestion plant, large amounts of solid substrate, such as biowaste with a high percentage of woody material, straw, hay or grass silage, can be processed, which in a wet digestion plant would cause flotation problems, etc. Thus, the question of flexibility in respect to wet and dry digestion plants cannot be answered in general, but only for specific cases and conditions.

If, however, a plant is being planned that will mainly be used to ferment liquid or semi-solid substrates, processes that use dry digestion will be ruled out as unsuitable, even though to a limited extent liquid substrates can be digested. Put simply, this means that it is the available range of substrates which determines the required process and plant type. Generalizations are difficult, because so much depends on the specific conditions of each case.

5.10. Substrate Heterogeneity and Its Influence on Process Stability

While wet digestion plants, as a rule, are operated with relatively homogenous substrate mixtures, plants dedicated to the various forms of dry digestion generally operate with less well mixed substrates. In extreme cases, such as the example of BEKON dry digestion, biowaste that has had no mechanical treatment at all, or had interfering materials (sand, foils, plastic bags, etc.) removed, can be digested (Figure 5.6), with no need for mechanical mixing during the digestion process, which is speeded up by prior inoculation of the fresh substrate with an equal quantity of digestion residue from the previous batch (Figure 5.7).

Figure 5.6. Biowaste from municipal waste collection of the City of Munich prior to the dry digestion process.

Figure 5.7. Simple mixing and inoculation of the substrate with digestion residue via a wheel loader (BEKON system). In larger plants, mixing and inoculation are carried out by using the same technology, but in a covered area.

With this method, there is the important question as to whether heterogeneity of the substrate in batch process can result in process instability and problems, particularly during the initial phase of digestion. It is particularly important to know whether the lack of comminution and homogenization, as well as the relatively small substrate surface area that results, leads to low specific yields of biogas. In addition, under some circumstances, problems can arise in establishing stable operating conditions when starting up such a biogas plant.

From the investigations described above and from general practical experience in starting up such biogas plants, one has to ask him/herself whether substrate heterogeneity, particularly in the batch process of dry digestion plants, is a problem.

5.11. Achieving Stable Operating Conditions in Dry Fermentation

In contrast to most other digestion processes, in the BEKON dry digestion process the digesters (reactors) are completely emptied after each cycle and subsequently refilled with a mixture of fresh substrate and inoculation material. It must be assumed that this causes a major disturbance of the physio-chemical conditions favorable to the digestion process. The entry of oxygen into the fermenter will raise the redox potential of the substrate to a level unsuitable for methane production, added to which, the optimal temperature must also reestablish itself. The heterogeneity of the substrate leads to local differences in conditions inside the digester. In order to investigate the starting up process and the amount of time it requires, the methane content of the biogas was measured during this initial phase of the digestion process. It is assumed that a methane content of more than 50 percent (by volume) indicates conditions favourable to methane production.

Figure 5.8 shows the methane content of biogas produced during the start-up phase for a series of digeston cycles. In all the digesters a biogas methane content of more than 30 percent was reached within 24 hours; within two days it had risen to over 50 percent. There were no digestion cycles in which favourable condition for methane production from biowaste was not reached within this period. The start-up phase of the BEKON dry digestion process was found to be stable despite (or perhaps because of) the substrate's high degree of heterogeneity.

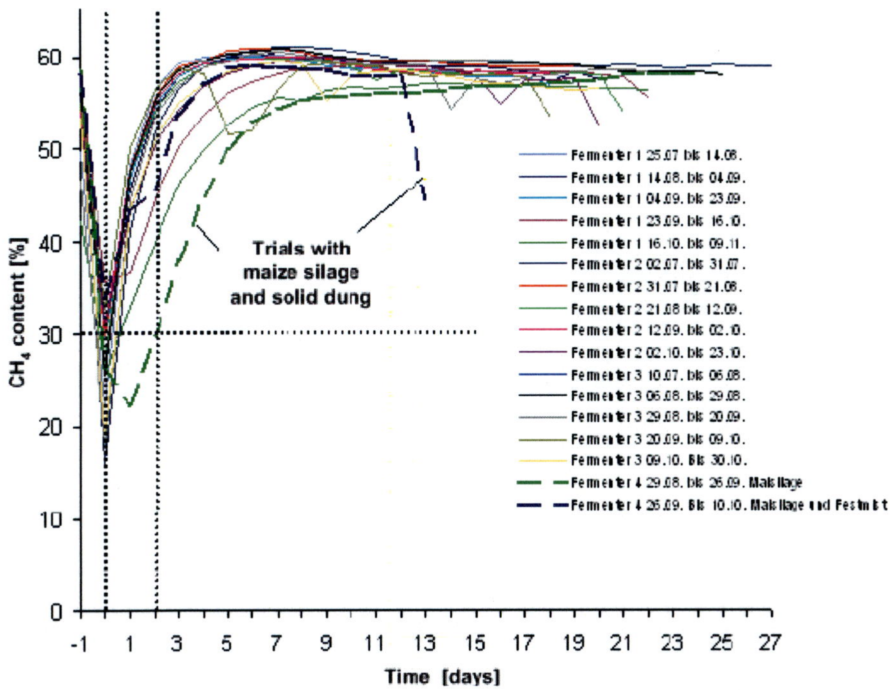

Figure 5.8. Increase in biogas methane content during the start-up phase of dry digestion.

The lack of homogeneity in the digester would seem to increase stability of the digestion process: when localized areas, thanks to a rich supply of readily degradable substrate, become strongly acidic, the acids responsible can be transported - by diffusion, or even more efficiently by being washed out through percolating liquor – to areas with conditions more favourable for their utilization by methane-producing bacteria. The substrate heterogeneity characteristic of the dry digestion process, favours the creation of micro-zones in which the different conditions necessary for the various microorganisms responsible for methane production can be better created than under the more homogenous conditions of wet digestion. Biological processes requiring different optimal conditions in respect to pH, organic acid concentration and hydrogen content can take place in close proximity to each other thanks to the formation of concentration gradients. Under favourable conditions, a kind of compartmentalized reactor is created through the process of dry digestion itself, which, depending on process control, facilitates various forms of material transformation. Thus, the oxygen, which enters the digester when it is freshly filled, is quickly consumed. Anaerobic conditions are created and digestion commences. The establishment of a methane producing biocenosis can be accelerated by inoculating the fresh substrate with digestion residue and/or through percolation of digestion liquor from a parallel operated, time-delayed digester.

The biogas used in most commercially available combined heat and power plants is generally recommended to have a methane content of at least 50 percent (by volume). This means that the gas produced in the first 24-48 hours following recharging of the digester cannot be used as it is to generate electricity. The energy content of a cubic metre of biogas with a methane content of about 30 percent (by volume) is about 3 kWh. However, when a number of digesters are operated, time-delayed, in parallel, it is possible to mix their biogas production, including that from the start-up phase, and thus ensure a combined methane content of over 50 percent for use in the generator.

5.12. Specific Biogas Yield from Dry Digestion

Since in dry digestion, as for example in the BEKON process, pretreatment of the substrate does not take place to any large degree, the specific biogas yield could be considerably lower than in wet digestion, where the substrate is subjected to extensive treatment prior to digestion. With this in mind, a material flow balance was produced for the discussed dry digestion plant (batch process, BEKON system). The duration of a digestion cycle in the case in question was 28 days. Through repeated inoculation, with already digested material from the preceding cycle, at the beginning of each new cycle, the retention time increased to a nominal 6-9 weeks, depending on inoculation ratio.

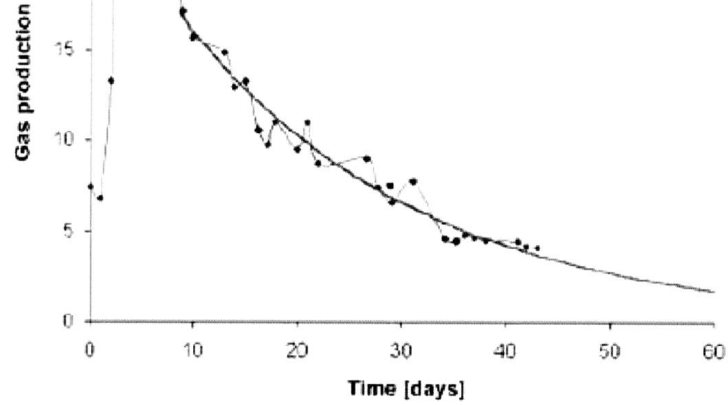

Figure 5.9. Gas production over a period of 43 days and extrapolated to 60 days (red curve).

In order to estimate the specific biogas yield of the biowaste it is necessary to differentiate between the biogas production from the fresh biowaste and from the inoculation material. The biogas yield from the inoculation material was determined through extension of the retention time (Figure 5.9). The rate of biogas production in one digester fluctuated in the case in question between 5-25 Nm^3/h. For the biowaste that was used, by a digestion cycle of 28 days, a specific biogas yield of 87 Nm^3/ton fresh weight was determined. The achieved specific biogas yield is comparable to that achieved using wet digestion. If the high proportion of structure-rich garden waste is taken into account (Figure 5.6), the specific biogas yield can be classified as good.

5.13. Content of Interferring Substances

All known digestion processes are sensitive – to varying degrees – to interfering substances. Exactly what substances are considered to be interfering depends very much on the technology employed and the purpose of treatment. Sand and stones, for example, are critical in wet digestion plants, where they cause increased wear and tear to machinery and form undesirable sediments in buffer tanks and digesters. In contrast, some dry digestion plants can tolerate virtually any amount of sand and stones without them causing any significant problems. The corresponding reduction in substrate would simply result in a reduced yield of biogas.

The need to remove interfering substances from the substrates to be treated, thus plays and important role in the choice of a particular digestion process. For the treatment of biowaste from MSW collection, two real-life examples will illustrate how different the procedures and results for wet and dry digestion can be.

In the wet digestion plant discussed previously (BVR Radeberg), up until 2004 biowaste from MSW collection was used as substrate. It is a single-stage wet digestion process capable of handling a large number of different solid, semi-solid and liquid substrates. For technical and processing reasons, the following materials are classified as interfering substances:

- Sand, glass and stones, because they result in sediment formation in intermediate storage tanks and digesters, and because their abrasiveness causes increased wear and tear on plant machinery (comminuators, pulpers, pumps, agitators, etc).
- Metals, because they can cause mechanical damage to, or blocking of, various plant components.
- Hard plastics and foils, because they can foul or block machinery or form floating layers.
- Woody components, because they can cause blockages and floating layers.

At this wet digestion plant considerable trouble had to be taken in respect to substrate preparation (comminution, removal of interfering substances; Figure 5.10), to ensure its smooth operation, and also, of course, to reduce the amount of interfering substances in the end product, which was secondary fertilizer.

Figure 5.10. Substrate preparation plant (comminuation and removal of interferring substances) for solid substrates handled by BVR (Bioverwertungsgesellschaft Radeberg GmbH).

In Figure 5.11 the ratio of interfering substances (coarse substances > 20 mm) to total biowaste for the years 2001 and 2002 is given. It is evident that quite a large proportion of the biowaste was removed as interfering substances. In 2001 the average proportion of interfering substances in collected biowaste (excluding sand and metals) was 25.9 %. To a large extent they consisted of woody biomass. Although such woody biomass would be quite acceptable in the production of secondary fertilizer through composting or dry digestion, it has to be removed from the substrate in wet digestion on account of the problems it can cause (floating layers, blockages, mechanical damage to machinery etc).

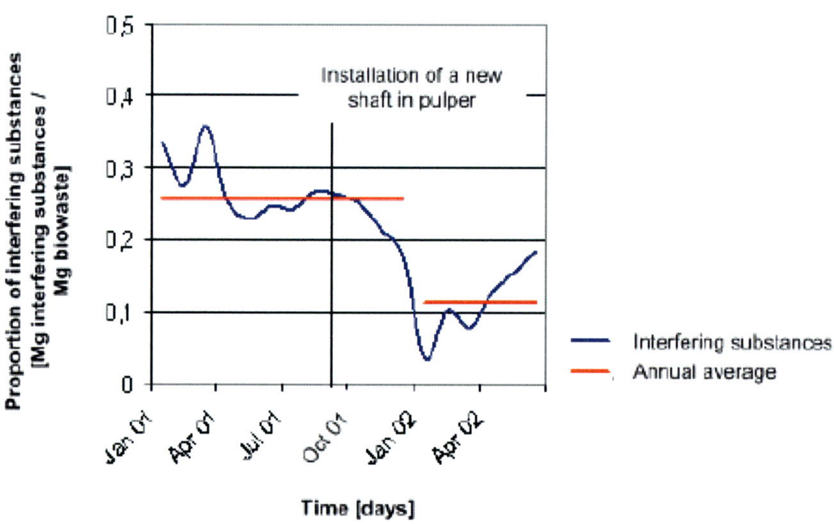

Figure 5.11. Proportion of interfering substances in biowaste treated by BGA Radeberg.

The amounts of interfering substances which have to be removed prior to wet digestion depend on the composition of the collected biowaste and on conditions in the treatment plant, particularly on the comminuting and pulping machinery. In the concrete case in question, the pulper was renewed in the course of regular servicing, which resulted in the proportion of interfering substances removed being reduced to about 12%. However, since plant machinery cannot be constantly renewed, the amount of interfering material (coarseness: > 20 mm) removed can be expected to be between 12-25%. In addition, 3-4% sand must be added to the proportion of interfering substances, which also has to be removed. The proportion of metals, which are removed via a Fe-separator, is negligible, although failure to remove them can result in mechanical damage to plant machinery. In contrast to the wet digestion plant described above, BEKON dry digestion plants are largely immune to the presence of interfering substances.

5.14. Own Electrical Energy Requirements

For the energy balance and thus also for the eco-efficiency of a digestion plant, its own energy requirements are of decisive importance. The proportion of the energy it produces required for a wet digestion plant are given in the literature, depending on plant size and the operated process, as between 15-70% of thermal energy and between 4-7% of electrical energy from combined heat and power generation through biogas production [1 to 10].

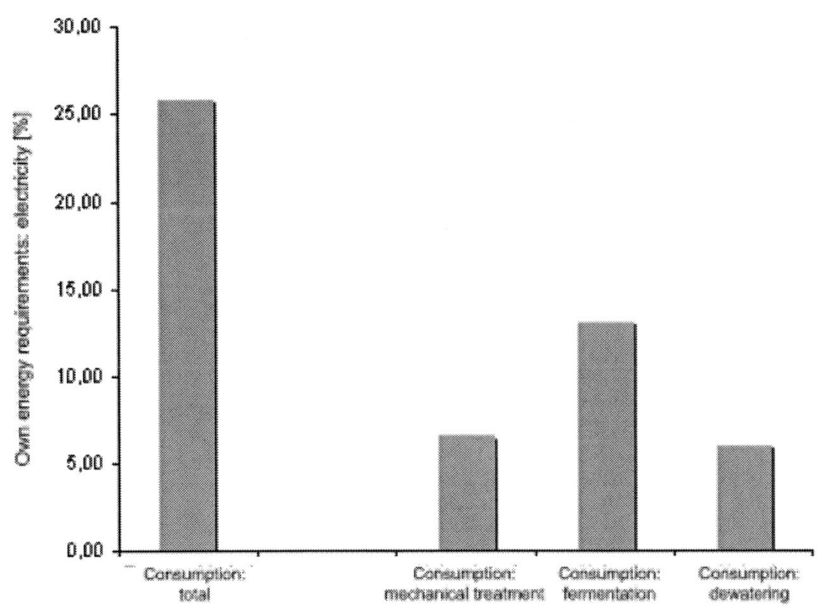

Figure 5.12. Own electrical energy requirements of the wet digestion plant (BVR Radeberg).

Other studies at the wet digestion plant, BVR Radeberg, show a much higher proportion of its electricity production being required for its own purposes, which in 2005 lay by 25.8 %. Its energy requirements are especially high on account of elaborate substrate preparation (comminution, slurrying, pressing of interfering substances, control of 9 storage tanks) and dewatering of the digestion residue. The energy requirements of just the core process of digestion amounted to about 12% of that produced (Figure 5.12). Although its own energy requirements, through the use of clarification sludge (lower specific biogas yield), tend to be higher than with substrates with a higher specific biogas yield, the example shows that the proportion of energy produced required by the wet digestion process itself can be considerable. In the dry digestion plant (BEKON, Munich) described above, plant energy requirements were also determined. In this plant there is no energy-intensive substrate preparation, nor is the substrate stirred in the digesters. Electrical energy is only needed for the percolator pumps, aerators, gas dryers, opening and closing of digester doors etc. Depending on plant size, the electricity requirements of the BEKON dry digestion lie at 2-3.5% of that produced. Thus, compared to wet digestion, dry digestion uses a considerably smaller proportion of its own electricity production.

REFEERENCES

Section 1

Bonomo A. (2003), *WTE Advances: The Experience of Brescia,* Presentation at the 11th North American Waste-to-Energy Conference, Tampa, FL, April 28-30.

Deriziotis P. and Themelis N.J. (2003), Substance and perceptions of environmental impacts of dioxin emissions, Proceedings of the 11th North American Waste-to-Energy Conference, *ASME International*, Tampa FL.

EAC (2000), Energy Answers Corporation, Albany, NY. (Gordon L. Sutin, private communications, plant visits by Earth Eng. Center personnel, company reports, Rochester MA, SEMASS plant operating data).

Energy Information Administration, *Electric Power Annual 2000*, DOE/EIA-0348 (2000)/1

Energy Information Administration/*Annual Energy Review 2001*, http://eia.doe.gov

European Union (1999), Council Directive 1999/31/EC of 26 April 1999 on the landfill of waste, *Official Journal of the European Communities*, L182/1-19.

Integrated Waste Services Association web site, http://www.wte.org/

Japanese Ministry of Environment (2000), Generation *and Management of MSW in 1999 in Japan*, Ministry of Environment, Tokyo, Japan.

Kaufman S.M., Goldstein N., Millrath K. and Themelis N.J. (2004), The State of Garbage in America, *BioCycle,* issue 45 (1), 31-41.

Kiser J.V.L. (2003), Recycling and Waste-to-Energy: The Ongoing Compatibility Success Story, *MSW Management*, issue 13 (4), 92-103.

Klein D.H. et al. (1975), Pathways of Thirty-seven Trace Elements through Coal-fired Power Plants, *Environmental Science and Technology*, issue 13, 455-459.

Roinen A. (1999), HSC Chemistry 4.0, *Outokumpu Research*, Outokumpu Oy, Pori, Finland.

Ruth A.L. (1998), Energy from Municipal Solid Waste: A comparison with coal combustion technology, *Prog. Energy Combustion Science*, issue 24, 545-564.

Scharff C. (2003), The waste site story - exploring the NIMBY syndrome, *Waste Management World*, issue 3 (3), 47-53.

SCS Engineers (1992), *New York City Waste Composition Study* (1989–1990), available from NYC Department of Sanitation.

Tchobanoglous G., Theisen H. and Vigil S. (1993), *Integrated Solid Waste Management*, McGraw-Hill, New York.

Themelis N.J. (2003), An Overview of the Global Waste-to-Energy Industry, *Waste Management World*, issue 3 (4), 40-47.

Themelis N.J. (2003), Status of the Global WTE Industry, *Waste Management World*, July-August.

Themelis N.J. and Gregory A. (2002), Mercury Emissions from High Temperature Sources in the Hudson Basin, *Proceedings of 10th North American Waste To Energy Conference*, Philadelphia, PA, May 6-8.

Themelis N.J. and Kim Y.H. (2002), Materials and Energy balances in a Large-Scale Aerobic Bioconversion Cell., *Journal of Waste Management and Research*, issue 20(3), 234–242.

Themelis N.J. and Todd C. (2004), Recycling in a Megacity, *Journal of Air and Waste Management Association* 54, accepted for publication.

US EPA (1993), Emission Factor Documentation for AP-42 Section 1.1 Bituminous and Sub-bituminous Coal Combustion, U.S. Environmental Protection Agency, Research Triangle Park, NC, April.

US EPA (1997), *Locating and Estimating Air Emissions from Sources of Dioxin and Furans*, (EPA-454/R-97-03; May)

US EPA (2003), Average Annual Emissions, *All Criteria Pollutants*, February.

US EPA (2003), *Municipal Solid Waste in The United States: 2001 Facts and Figures*, US EPA, Office of Solid Waste and Emergency Response, Washington, D.C.

Section 2

Bilitewski B., Schirmer M., Niestroj J., Wagner J., Müller H. (2005), Ökologische Effekte der Müllverbrennung durch Energienutzung, Dokumentation und Forschungsbericht der EdDE e.V. Entsorgergemeinschaft der Deutschen Entsorgungswirtschaft Köln, Heft Nr. 10, Eigenverlag, 12/2005 (in German).

Vehlow J. (2006), Internationale Entwicklung der thermischen Abfallbehandlung in 11. Fachtagung Thermische Abfallbehandlung, 14.-15.3.2006 in München, Schriftenreihe des Fachgebietes Abfalltechnik, Universität Kassel (in German).

Section 3

Bundesamt für Umwelt, Wald und Landschaft (Buwal) (1998), *Entsorgung von Abfällen in Zementwerken*. (Richtlinie), Bern.

Eckardt S. (2005), Anforderungen an die Aufbereitung von Siedlungs- und Produktionsabfällen zu Ersatzbrennstoffen für die thermische Nutzung in Kraftwerken und industriellen Feuerungsanlagen (Dissertation), Beiträge zu Abfallwirtschaft/Atlasten, Band 41, *Forum für Abfallwirtschaft und Altlasten*, Pirna (In German).

Keldenich K. and Marzi Th. (2006), *European Study on RDF*, Frauenhofer Institute for Environmental Safety and Energy Technology Umsicht.

Ketelsen K. (1999), Aufbereitungstechnologien für Sekundärbrennstoffe: Technik, Ökologie, Kosten, Gallenkemper et. al. [Hrsg.]: 6. *Münsteraner Abfallwirtschaftstage* (Tagungsband), Münster.

LAGA (1997), Maßstäbe und Kriterien für die energetische Verwertung von Abfällen in Zementwerken. Entwurf vom 7.10.97.

Püchel H.J. (1999), Die Herstellung von Substitutbrennstoffen aus Restmüll und ihre Qualitätssicherung, SIDAF- [Hrsg.], Kostenminimierte Abfallentsorgung durch Verfahrenskombination Kalt/Warm. Freiberg.

Schirmer M. (2007), Analytik und Freisetzungsverhalten von Chlor in abfallstämmigen Brennstoffen (Dissertation), Beiträge zu Abfallwirtschaft/Atlasten, Band 47, *Forum für Abfallwirtschaft und Altlasten*, Pirna (in German).

UBA (1998), Umweltbundesamt, Fachbereich III 2.2: Daten und Fakten: Kohlendioxidemissionen. October.

Section 4

Bilitewski B., Härdtle G., Marek K. and Weissbach A. (1997), *Waste Management*, Springer Verlag, Heidelberg.

Rasch R. (1977), Gaserzeuger-Praxis und Müllvergasung. *Aufbereitungstechnik*, issue 3, 138-143.

Thomanetz E. (1983), Möglichkeiten und Grenzen der Erzeugung von Energie und Rohstoffen durch die Pyrolyse von Abfällen, 16. AWS, TU Berlin, 69-91

Section 5

Bundesforschungsanstalt für Landwirtschaft (FAL) (2005), Ergebnisse des Biogas-Messprogramms, *Fachagentur für Nachwachsende Rohstoffe* e.V. (FNR) (Hrsg.), Gülzow.

Chandler J.A., Jewell W.J., Gossett J.M., Van Soest P.J. and Robertson J.B. (1980), Predicting methane fermentation biodegradability, *Biotechnology and Bioengineering Symposium*, issue 10, 93-107.

Dalheimer F., Heerenklage J. and Stegmann R. (1999), A multichamber anaerobic dry fermentation plant for the pretreatment of MSW, in: Cossu, Christensen, Stegmann (Hrsg.): *Proceedings of Sardinia '99, International Symposium*, Vol. I, Cagliari, 443-446.

Keymer U. (2006), Gewinn schwankt schnell um 35000 Euro, top agrar Jahrbuch neue Energien 2006, Münster-Hiltrup.

Kottner M. (2002), Dry Fermentation-A new method for biological treatment. *International Biogas and Bioenergy Centre of Competence (IBBC).* Jagst, Germany, 16.

Lutz P. and Wittmaier M. (2005), Praxiserfahrungen mit Bioabfallen aus der Getrenntsammlung. *Mull und Abfall* 2 - 05, 91-93.

Schäfer J. (1998), Verfahrenstechnische Untersuchungen zur Vergärung von Biomüll und Klärschlamm, Berichte aus der Forschung Nr. 012,Fraunhofer IRB Verlag, Stuttgart.

Schüsseler P. (2006), Ergebnisse der Diskussion und Zusammenfassung in: Trockenfermentation - Stand der Entwicklungen und weiterer F+E-Bedarf, *Gülzower Fachgespräche*, Bd 24.

Weiland P. (2006), Stand der Technik bei der Trockenfermentation – Aktuelle Entwicklungen in: Trockenfermentation - Stand der Entwicklungen und weiterer F+E-Bedarf, *Gülzower Fachgespräche*, Bd 24.

In: Solid Waste: New Research
Editors: A. S. Martin and J. D. Suarez
SBN: 978-1-60456-809-7

Chapter 2

SOLID WASTE MANAGEMENT, RECENT TRENDS AND CURRENT PRACTICES FOR SECONDARY PROCESSING OF ZINC AND LEAD INDUSTRIES IN INDIA

Archana Agrawal* and K. K. Sahu
Metal Extraction and Forming Division, National Metallurgical Laboratory,
Jamshedpur, India

ABSTRACT

Almost all the metallurgical processes are associated with the generation of wastes/residues which may be hazardous or non-hazardous in nature depending upon the criteria specified by institutions like US EPA etc. The wastes containing heavy and toxic metals such as arsenic, cadmium, chromium, nickel, lead, copper, mercury, zinc etc. are present beyond permissible limits deemed to be disposed off. Due to the implementation of stricter environmental laws and economic reasons all the metallurgical industries are now forced to go for eco-friendly technologies to produce metal and other related products world over. However, generation of wastes is the integral part of metallurgical industries which can not be ruled out, therefore if the wastes/residues are hazardous in nature they generally have to be treated or disposed off in safe and designated dumping sites. If these wastes/residues are non-hazardous in nature then they may be suitably used as secondary raw material for the recovery of metals which are in growing demand all over the world. Zinc is in growing demand all over the world. In India a major amount of zinc is imported and therefore processing of zinc secondaries will supplement in satisfying the gap between demand and supply to some extent. Similarly processing of lead secondaries is important because of their relative high metal content, besides low energy and cost involved in recovering the metal. This chapter highlights the production capacity, type and quantity of solid wastes generated, their chemical composition and treatment/disposal options for the Indian lead and zinc industries. Zinc tailing, slag, leach residue, jarosite residue, β-cake, etc. from zinc industries and BF slag, flue dust, ISF slag etc. from lead industries are the major solid waste generated from various process and needs attention. Although all the metal producing industries in organised sector are now

* Author for correspondence e mail address archana@nmlindia.org or archana_nml03@yahoo.com

taking care of the environment and waste management related problems, but pollution from unorganized lead units are the major cause of concern. Permissible limits of toxic constituents in zinc based secondaries and threshold zinc concentration for both indigenous and imported raw materials were worked out at National Metallurgical Laboratory (NML). An overview of the current practices and recent trends in the secondary processing of zinc and lead and the attempts made to recycle/recover metal values and production of value added products, are discussed in the text. Various processes, particularly hydrometallurgical ones, already developed or in the developmental stages, are discussed. Attempts made by various laboratories and industries towards the development of eco-friendly processes for the recovery of zinc and lead from secondary raw material are also described.

Key words: pollution measures, recycling, secondary zinc, lead, lead acid battery

INTRODUCTION

Metallurgical industries generate vast quantities of solid wastes such as slag, ash, sludge, dross, grindings, turnings, clippings, residues and secondaries. During the last few decades rapid industrialization has led to many fold decline in the quality of environment. Some of the solid wastes produced during metal extraction are hazardous in nature due to the presence of certain toxic metals which contaminate the surface and ground water through the leachate generated at the dump-sites, posing a risk to the life of the living organism. Effects of pollution due to the toxic constituents of the above leachate, are usually noticed in the long run. It is for this reason that the industries are not much concerned about the solid metallic wastes that are invariably thrown and dumped unsystematically during metal production. The type of metallic waste generated, depends on the quality of raw materials used, the process and the treatment methods. Irrespective of the physical forms such as solid, liquid and gas, the wastes generated are eventually of two types: hazardous and nonhazardous. Amongst the heavy and toxic metals, arsenic, cadmium, chromium, nickel, lead, copper, mercury and zinc are considered deleterious to the environment, when their concentrations are more than the stipulated limits. The major environmental degradation is caused by copper, lead and zinc industries, often using imported zinc ash/residue, dross, skimming, lead scrap etc. It is therefore considered a major problem for allowing to process such material with out proper environment friendly technology and safety measures

This chapter deals with an overview on the quantity and composition of the various solid wastes/ secondaries generated, processing options for such materials and their management and the current trends and recent practice in the utilization of various wastes containing zinc and lead. Zinc is the most used non-ferrous metals next to aluminum and copper. Apart from its major applications (about 70%) for galvanizing of steel, the other important areas of applications are as sheets, anodes, casting, chemicals, micronutrients, paints, dry cells etc. As the usage of zinc increases, the gap between demand and supply is widening [1,2], the possibilities of recycling of zinc exist from many of these applications. Out of about 10 million tons of zinc production globally about 30% of zinc is produced from the secondary sources.

Major raw materials available for secondary zinc units in India are zinc ash, dross and blowings from galvanising plants, brass waste such as brass dross/ash and scrap, electric arc furnace (EAF) dust, die-cast scrap, sheet cutting, old zinc roofing and anodes, skimmings and residues, pickle salt wastes etc. Since the indigenous availability of the above secondary raw materials is very limited these have to be imported from other countries. It was found that many of these imported secondaries were toxic in nature due to the presence of toxic radicals more than the permissible limits designated by US EPA [3]. Hence Basal Ban was implemented on the import (trans boundary movement) [4,5] of these secondary raw materials because of which secondary zinc industries had to face a major setback. Many of the industries in the organised and unorganised sectors were closed for the want of raw material. Both pyrometallurgical as well as hydrometallurgical processes are adopted worldwide, whereas mostly hydrometallurgical processes are being practised in India to treat secondaries, which comprises of leaching, iron removal, purification and electrowinning. The generation of zinc-based secondaries in our country is limited, whereas the zinc based processing industries have grown over the years in order to meet the demand of zinc metal and its compounds. Since processing of secondary zinc materials gives better returns, there has been a tendency to operate such industries in medium and small scale sectors using both the indigenous and imported materials. After the implementation of strict environmental rules all those secondary units which have opted for eco-friendly technology and have proper pollution control devices have applied for the registration and import licence to Ministry of Environment and forest and a few of them have got the clearance and come back to life. Thus an attempt has been made to focus upon the newer technologies for the secondary processing of zinc.

Similarly with the increase in the usage of lead, the gap between demand and supply is widening [1]. In terms of worldwide annual tonnage consumption, lead is fifth important metal in use. The current demand for lead in India is about 1,61,000 TPA, of which about 50% is met through imports. Spent battery scraps are the source of 80% lead that is recycled in the country and the balance is imported [6]. Due to the depletion of the primary source for lead, primary production units of lead can never bridge this gap and therefore the concept of secondary metal production by using wastes/residues from primary metallurgical processing of metals, alloys etc., will help in supplementing the primary metal production. Raw materials for secondary lead production units are mainly spent lead-acid batteries, used scrap like worn-out sheets, pipes, solder, dusts, drosses etc. The advantage of processing secondaries is the conservation of energy, which is far richer in metal content than the primary resource. An increasing number of eco-friendly technologies are available for the production of secondary metals. Major primary metal producing units and secondary producers in organised sectors already have the proper pollution control devices installed however, the main problems faced in production is simply the unavailability of the secondaries to them. Small scale secondary unit in unorganised sector, devoid of ecofriendly technology, uses lead acid battery as raw material causes maximum pollution. Newer technologies for the secondary processing of lead [7] are also given keeping all the above factors in mind. Some efforts made by different laboratory in India in this area are also discussed.

ZINC INDUSTRIES

Earlier the installed smelting capacity for zinc in India was 199,000 tonnes of which HZL had 169,000 tonnes (Debari: 59,000 tonnes, Visakhapatnam: 40,000 tonnes, and Chanderia: 70,000 tonnes). After being taken over by Sterlite group, HZL has raised its production capacity to 220,000 tonnes a year by removing bottlenecks and improving efficiency [8]. Recently HZL has further expanded the capacity to 400,000 tonnes [9]. The company estimated to produced about 300,000 tonnes (including 230,000tonns though domestic capacity and 70,000 tonnes through toll processing route) in the year 2004-05 against the actual output of 260,000 tonnes in 2003-03 (including 225,000 tonnes through domestic capacity and 35,000 tonne through toll processing route). Binani Industries Ltd (BZL) is estimated to produce 35,000 tonnes for 2004-05 against about 32,000 tonnes in 2003-04. Thus the total supply by these two industries is about 335,000 tonnes for 2004-05 against 290,000 tonnes in 2003-04 [10]. However the demand and supply gap would be around 65,000 tonnes for 2004-05 against 60,000 tonnes in 2003-04. The demand for refined zinc is 400,000 tonnes in the fiscal year of 2004-05 which is growing at a rate of 12-15% annually and likely to sustain for next few years [9]. HZL uses mainly indigenous zinc concentrate and ocassionaly imports from other countries while Binani Ltd. uses imported concentrates only. About 15-20% of zinc demand in India is met through secondary production against world average of 32-35%. There are about 40 secondary units and more that 200 zinc chemicals units (zinc oxide, zinc sulphide, zinc chloride etc). These play an important role in balancing the demand-supply gap in the country. The recyclers depend on the import to keep their units running economically since local wastes are insufficient to cater the demand.

All zinc wastes containing zinc have been categorised as hazardous waste under the Basel convention. Some selected dross, zinc scrap and skimming are in green list of waste and zinc ash happens to be in amber list. The European Union has placed all zinc containing wastes in the list of hazardous wastes. Because of the likely contaminants present in the zinc skimmings, a product of galvanising operation, they have been placed under the categories of wastes for which the level of contamination is essential to quantify. The residue generated from these materials by secondary units may have hazardous character depending on the concentration of the toxic constituents.

Waste Treatment/Disposal Practices in Zinc Industries

The details of waste generated by some of the major zinc plants, their quantity, composition, and treatment/disposal practices are incorporated in Table 1. At Binani Zinc smelter[11], the jarosite residue generated is pumped out and safely stored inside the company premises in a large pond of 1,40,000 M^3 capacity. The residue contains toxic constituents in good amount like 3-5% Zn, 8-10% Pb, and 40-50% SO_4. Ground water monitoring is periodically conducted from the tube wells installed around the ponds. Steps have been initiated for the construction of new jarosite ponds by conducting Rapid Environmental Impact Assessment (REIA). Based on the recommendation submitted by the public hearing panel, the government of Kerela notified the site for construction of jarosite pond. The design of the proposed pond is as per the guidelines of United States Environmental Protection Agency (USEPA) and Ministry of Environment and Forest (MoEF) for secure landfills. The pond design is based on geochemical investigations and compatibility studies with the waste. The pond is lined on all the sides and bottom with two layers of LDPE sheets of 250 micron thickness, heat sealed and protected by burnt clay bricks. The primary and secondary cake leach residues are also stored inside company premises in a covered shed provided with acid proof bricks at the bottom and disposed by way of sales.

Table 1. Details of waste generated and treatment/disposal practices in zinc plant.

Sl. No	Name of the Industry	Waste Generated: Quantity & Composition (%)	Waste Reprocessing/ Dumping
1	Binani Zinc	**-Jarosite residue**:	-Stored inside the premises in a large
	Kerala[10]	Zn: 3-5%, Fe: 20-28%	pond of 14,000 M^3 lined with LPDE
		Pb: 8-10%, SO_4: 40-46%	sheets
		-Primary cake leach residue	- Both Primary and secondary Leach
		Zn: 2-4%, Cd: 0.5-2%,	residue stored inside the premises in
		Cu: 40-50%, Co: 0.02-0.03%	a covered shed provided with acid
2	HZL	**-Secondary leach residue**	proof bricks at the bottom
	Chanderia	Zn: 20-30%, Cd: 1.5%, Cu: 10-15%, Co: 0.2-.5%	
	Rajasthan[11-13]	**-ISF Slag** - 42, 000 t/y	-Dumped & commercial disposal
		Zn: 6-9, Pb: 0.6-0.7, Cu: 0.19, Sb: .01, Ag: 19	-Goes for Cadmium recovery.
3	Debari Smelter	g/MT, Au: 0.04 g/t, FeO: 33	
	plant, HZL,	**-Effluent plant sludge**	-Goes for Ag recovery
	Rajasthan[10, 14-16]	(Details of other wastes covered in lead section)	
		-Drum Cake	
		Zn: 17, Fe: 32, Pb: 5.4, S: 5.4, Cd: 0.25, Cu: 0.11,	-Goes for Cd recovery
		Co: 0.003, Ni: 0.005, MgO: 0.6, Mn: 2.1, Sb:	-For Co recovery (Mostly stored)
		0.004, As: 0.05, CaO: 3.0, SiO_2: 3.9, Ag: 0.038	
		-Cadmium cake	-Dumping
		Zn: 36, Cd: 14.0, Cu: 2.2,	-Dumping
		-Beta Cake	
4	Zinc Lead Smelter	Zn: 9.16, Cu: 0.05, Cd: 0.5, Mn: 2.6, Co: 1.45,	-Transferred to Waelz Kiln for zinc
	plant, HZL,	Ni: 0.31, Fe: 3.33	recovery
	Visakhapat-nam[10, 13,14]	**-Jarosite residue**: 41,000 t/y	as zinc oxide
		Zn: 3-4, Fe: 19-20, Pb: 3-4, Cu: 0.04, Cd: 0.07,	-Goes to klinker klin for separation
		CaO: 12.0, SiO_2: 4.0, Al_2O_3: 4.4,	of zinc cadmium & lead and leaching
		-Zinc Tailing	of zinc in main unit.
5	Sunrise Zinc	Zn: 16.3, Fe: 29.5, Pb: 7.0, Cu: 0.31, Cd: 0.26,	
	Cuncolim,	Ag: 0.013, MgO: 0.8, CaO: 6.4, SiO_2: 5.3, Al_2O_3:	
	GOA[15,17]	5.9, S: 0.36	-Waste is disposed off in a secured
		-Moore cake	land fill
		Pb: 5.6-5.8, Fe: 22-25, Zn: 20-22, Cd: 0.05-0.075,	-Secured land fill
6		Ag: 0.0125, Ca: 3-4.	
	Bharat Zinc[10]	**-Raw Zinc Oxide**: 6,950 t/y	
		Zn: 55 - 60, .Pb: 7-10, Ag: 60ppm, Cu: 0.01-0.03,	
		As: 0.03, Sb: 0.01-0.03, F: 0.05-0.09, C: 1-2,	
		FeO: 1.0, Cd: 0.28, CaO: 0.2, MgO: 0.2, Al_2O_3:	
		0.8-1.0	
		-Sintered Zinc Oxide: 6950 t/y	
		Zn: 65, Pb: 8-9, Cd: 0.02	
		-Clinker Product	
		Zn: 58-62, Pb: 5-7, SiO_2: 5.0, Fe: 3-5, Cd: 0.01-0.015, Ag: 0.0042	
		Sludge - 2,400 t/y	
		Zn: 1.26, Cu: 1.65, Fe: 2.94, Al:0.6, Pb:1.42, Si: 3.8, Gypsum: 24.5	
		Filter Cake	

HZL has been in forefront of adopting appropriate pollution control and environmental protection measures in all its operations. Environment Impact Assessments are undertaken for new projects to formulate Environment Management Plans for mitigating any possible environmental impacts. The environmental measures at mining units of HZL include provision for dust suppression and collection system, disposal of beneficiation plant tailing into tailing ponds, recycling of tailing dam water to beneficiation plants as process water so as to maintain zero discharge, green belt development to improve landscape, to work as dust sinks and achieve noise attenuation. At its open pit Rampura Agucha Mine, a number of measures have been taken to ensure minimum ground vibrations due to blasting, and each and every blast is monitored for the vibration levels.

At the smelting units, the environmental protection measures include sulphuric acid plants on Double Conversion Double Absorption (DCDA) technology to ensure minimum sulphur di-oxide emissions. Single Conversion Single Absorption (SCSA) Sulphuric Acid plants also have been provided with Tail Gas Treatment facilities for ensuring minimum sulphur dioxide emissions. In addition, other elaborate dust collection and gas cleaning facilities have been provided to ensure minimized gaseous emissions. Integrated effluent treatment plants have been provided to ensure that treated effluents meet all the criteria stipulated by the Central Pollution Control Boards.

The waste generated as tailing in the concentrator plant at HZL though not included in this article is mostly dumped in the specific area, besides the back filling and in the tailing dams and also used in constructing embankment, mine road, play ground, filling low lying area etc. The major wastes generated from the smelters are the leach residues and the jarosite residues. The residue treatment plant at Debari, HZL is in operation to treat and recover zinc and other metal values. The final waste, jarosite cake is disposed off in storage pond after lime addition to render it inert. The intermediate residues like drum cake and cadmium containing cake are processed for the recovery of Ag and Cd in these units. β-cake is another material consisting of cobalt, and is mostly stored for Co recovery [16]. The process used for Co recovery involves fluidized bed roasting of β-cake followed by H_2SO_4 leaching, solution purification, solvent extraction and electrowinning. The pilot plant installed at Debari (HZL), Udaipur for Co recovery is not utilized for continuous production of this metal. Maximum recovery of cobalt from β-cake is reported to be about 60%.

The imperial smelter furnace (ISF) slag generated at Chanderia unit is mostly stored in a separate dump yard with minor commercial disposal. Though the slag is vitreous in nature the toxic constituents are in sufficient amount and may prove to be hazardous due to mobility of the ions by weathering action.

Manufacturers of the secondary zinc use indigenously available secondaries and imported materials (Zn ash/dross/skimming, EAF dust). Two types of dross namely top dross that floats on the top of the bath and the bottom dross that sinks to the bottom of the galvanising bath based on the specific gravity of the material, are obtained. They account for 10-20% loss of Zn during galvanising. Zinc content in drosses lies in the range 93-98% (Table 2).

Table 2. Zinc waste from galvanising sectors

Type of galvanising	Zinc consumption (T)	25% zinc waste (T)
Sheet galvanising	132500	33000
Pipe galvanising	79500	19800
General galvanising	39750	9900
Wire galvanising	13250	3300
Total	265000	66000

Quite often the technology followed is hydrometallurgical based involving leaching, metal purification, separation, precipitation and electrolysis. In some of the units, ZnO is manufactured from the secondary zinc following pyrometallurgical processes, which involves carbon reduction and vaporisation of zinc followed by controlled oxidation to produce ZnO. Waste is mostly in the form of residues, which are often disposed off as unsystematic landfills, though some industries follow the safe handling and disposal procedure laid down officially by monitoring agencies.

Over the years zinc sulphate industry dealing with micronutrients has developed technology for handling air, water and solid wastes[19]. The sludge obtained in the process from the filter press is washed to recover both water soluble and acid soluble zinc and washing is then recycled. This sludge is given a treatment of excess lime, which converts water-soluble zinc to insoluble form. Lead that is present originally in the zinc ash (0.3 – 0.8%) will also report in the sludge as insoluble form of lead metal/oxide or lead sulphate. To ensure that treated sludge does not carry any soluble salts of heavy metal ions, it is kept for 10-15 days for curing before final disposal. The treated sludge is disposed off which may cause some damage to the environment in the long run unless Environmental Impact Analysis (EIA) finds it otherwise.

Secondary Zinc Processing

In India about 70% of the total zinc produced is used in galvanising sector of which 25% of zinc is lost as zinc waste (Table 2) [20] Apart from its major applications to the tune of about 75% as a protection material, the other important areas of applications for zinc and zinc alloys are as alloys and dye casting 15%, dry batteries 10%, chemicals, fertilizer and micronutrients, paints etc. 5%. Possibilities of recycling of zinc exist from most of these applications. India is a net importer of zinc and with rising demand there has been increasing interest in secondary zinc production. Recycling of zinc offers both environmental and economic benefits by: (i) reducing energy required for mining and processing (ii) reducing volume of material that end up on land filling (iii) relieving environmental impacts on the land and water and (iv) conserving zinc ores.

The important sources of secondary zinc available for recycling are: pure zinc scrap in the form of sheet cuttings, zinc roofing, old zinc anodes etc., alloys containing zinc as a major constituent such as die casting alloys, brass dross/ash and scrap, other scrap alloys, zinc rich residues like zinc dross, zinc ash/skimmings and blowings from galvanising industries. The normal range of constituent in galvanising residues and other zinc sources are given in Table 3.

Table 3. Range of constituents in secondary zinc sources

Secondary Zn Source	Industry	Composition(%)					
		Zn	Cu	Cd	Fe	Cl	Others
Zinc dross (Bottom)	Galvanising	93-98	0.0-0.1	0-0.2	2-5	0-0.1	Pb 0.0-0.5 Sn: 0-0.2 Al: 0.5-4.0
Zinc dross (Top)	Galvanising	93-98	0.0-0.1	-	0.5-3.0	0-0.1	Pb 0.3-2.5 Sn: 0-0.2 Al: 0-1.5
Zinc ash	Galvanising	Zn(T) 80-90 Zn(metallic) 40-60	-	-	0.5-1.0	2.0-4.0	Pb:0.2-1.25 Al: 0.0-0.3
Zinc Skimming	Galvanising	50-60		0-0.2	0 - 1	10 - 25	NH_4Cl - 3.1
Brass dross	Brass foundries	41-42	43-45	-	1-2	0.1-0.5	SiO_2 1-2
Brass ash	Brass foundries	40-50	15-20		1.0-2.0		SiO_2:0.25-20, Al:0.0-0.5, Pb: 0.2-1.0
Copper cement	Zinc plant	6.0-9.5	5.5-7.0	-	1-3	-	-
Flue dust	Pb & Cu smelters	30-32	5-6	-	1.5-2	7-8	-
Oxide dust	Waelz process	60-65	0.1-0.3	-	1-1.5	1.5-2	-
Steel fumes	Steel furnace	30-30.5	22-24	-	0.1-0.3	1.5-2	-

World over the contribution [2] from different sources of zinc for recycling include:

Brass scrap	: 32%	Zinc sheet	: 10%
Die casting scrap	: 16%	Steel plant dust	: 8%
Galvanising residue	: 23%	Others	: 11%

There are about 40 secondary zinc units in India having a capacity of about 60,000 tonnes but actually producing 30,000 - 35,000 tonnes play a balance role in bridging the demand-supply gap . Therefore the secondary raw material has to be imported but following Basel convention's 1989 decision where several developing countries put a complete ban on imports of hazardous and toxic wastes in to their territory. Among the items banned were lead and zinc ash and their skimming which has resulted in closure of most of the secondary units manufacturing zinc metal from scrap and wastes and consequently the country has to import an extra amount of about 35,000 tonnes per year (t/y) of zinc.

Both pyro- as well as hydrometallurgical processes are practised world wide to treat zinc wastes. There are various pyro-metallurgical processes depending on the zinc concentration in the secondary feed, such as

- Electrothermic process for high and low zinc contained residues
- Imperial Smelting Furnace (ISF) process for medium zinc contained [1]

Waelz kiln or QSL process, Larvic process, New Jersey's continuous vertical retort zinc distillation process, SKF plasma process [21], Mintek Enviroplas [22, 23] are used world wide. Waelz kiln process is most extensively applied in the pyrometallurgical treatment of residues. About 76% of total electric arc furnace (EAF) dust is treated by Waelz kiln in USA, Japan and Europe.

Zinc dross is a metallic alloy of zinc and some iron which is generated during galvanizing operation is often processed by pyrometallurgical techniques. The commonly practiced methods for zinc dross processing [24] are:

Liquidation and Remelting

In this process the material is kept on a sloping hearth at a suitable temperature so that the molten zinc trickles down without effecting iron. However liquation is not suitable for the production of high purity zinc for galvanizing since the zinc that trickles down carries fine dross crystals and is oxidized in the process. The working temperature is high as result zinc is contaminated with iron due to increased solubility. Recovery of zinc by this process is around 30% of the weight of dross treated. It has been reported that higher recovery of zinc from dross could be achieved by centrifuging and squeezing process.

Aluminum Process

Based on the aluminum addition Schmidt developed method for the recovery of zinc.

Addition of aluminum leads to the displacement of zinc from crystals of zeta phases ($FeZn_{13}$) as the affinity of Al for Fe is greater than that for Zn. The alloy of iron and Al($FeAl_3$) requires only 1.5 parts of Al to 1 part of iron compared to the 15.5 parts of Zn in zeta phase. Density of $FeAl_3$ is much lighter than molten zinc, hence it floats on the surface and its skimming is easier and efficient than scooping from the bottom of the bath. The operation is carried out at about 720 °C and after the removal of skimming the metal is cooled and casted. Recovery of metal is around 85%. This process is feasible at small scale also.

Electro Thermal Process

This process of recovery of zinc has a number of advantages over the indirect distillation method such as:

- Electricity is used as a source of heating making the process energy efficient
- Recovery of zinc is 98% against 92% in distillation method
- Removal of solid residue left after distillation is easier because as the crust can be remelted and tapped out by raising the temperature making the temperature continuous one.
- No retort or pot is required for holding the dross for distillation.

The process is feasible where the electricity is cheap. The capital investment is large and is feasible on large scale.

Distillation Process

In this process zinc from dross is distilled off as zinc vapour from a retort and condensed in a suitable condenser. The process requires specialized equipments and operates at high temperature. The recovery and purity of zinc is high. Three types of furnaces namely stationary, horizontal and tilting type are commonly used. The stationary and horizontal furnaces are similar in construction and size. These are made of firebricks and fired with gas or oil. The tilting type furnace consists of a large cylindrical steel casing lined with firebricks and mounted on triunions, which facilitated tapping at an angle. In these furnaces retort is used for melting the dross. The metal vapour/fumes from the retort passes into the condenser where it is cooled either as metal or as dust depending on the design of the condenser, alternatively the fumes/vapours from retort is allowed to burn in the adjoining combustion chamber in presence of air to form zinc oxide.

In India, the use of pyrometallurgical techniques for the recycling of zinc waste has been of less significant. Mostly hydrometallurgical processes are practiced in India. Hydrometallurgical methods are comparatively cleaner and can be adopted in small and medium scale industries. The three main hydrometallurgical methods for treating the secondaries such as zinc ash/skimming, brass ashes etc. are

- Electrowinning process,
- Solvent Extraction-Electrowinning process
- Crystallization process for manufacture of zinc sulphate,

Electrowinning process: The process generally uses zinc ash and other raw materials containing low level of impurities such as copper, cadmium etc. The processing steps mainly consist of material preparation, leaching and purification, electrowinning and melting. The material is crushed and pulverised to separate metallic zinc from fine ash. GOB (Good Ordinary Brand, 98.5% pure) zinc is obtained on melting and casting the separated metallic zinc.

Fine ash is subjected to calcination to remove chlorides and pulverised again to get particle size of <100 mesh. The calcined fine ash is treated with sulphuric acid/spent electrolyte to get zinc sulphate solution as per equation 1.

$$ZnO + H_2SO_4 \longrightarrow ZnSO_4 + H_2O \qquad (1)$$

During leaching, compressed air is passed and pyrolusite (MnO_2) is added to oxidise ferrous iron to ferric state. Leaching is continued till the pH of the slurry reaches to 4.5-5. At this pH, ferric iron is hydrolysed to ferric hydroxide and precipitated along with other minor impurities such as Al and partially As, Sb, etc. A suitable flocculent is added to the slurry and discharged to a thickener for solid liquid separation. The solid filter cake generated is sent for disposal after suitable treatment and the filtrate containing concentrated zinc sulphate along with minor impurities like Cu, Cd, Co, Ni, As, Sb, Ge etc. goes for three stages of purification.

- First stage purification removes copper and cadmium as cement with the addition of zinc dust/powder at ambient temperature.

- Second stage of purification is done by cementation with zinc dust/powder at higher temperature of about 85 °C (hot purification) in presence of potassium antimony tartrate, which practically removes all the impurities. Alternatively, a few industries adopt sometimes-chemical precipitation method using di-methyl glyoxime (DMG) for Ni and β-naphthol for cobalt removal.
- In the third and final stage purification step, activated charcoal is added to the purified solution to absorb unwanted organics and other impurities. This step is also considered as the polishing stage.

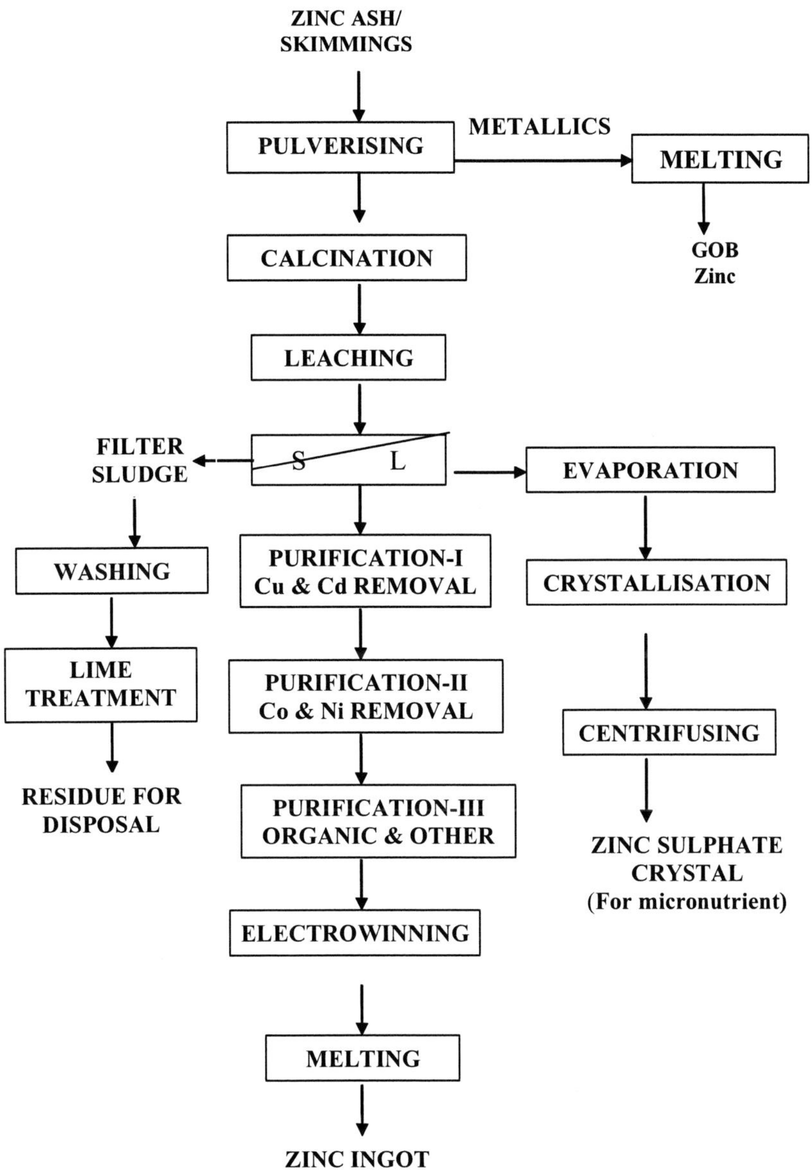

Figure 1. General flow sheet for producing electrolytic zinc metal and micronutrient from zinc ash and skimmings.

The purified solution is cooled to a temperature of 35-40 °C and is electrolysed in an electrolytic cell using lead anodes and aluminium cathodes. The zinc metal is deposited at the cathode following the reaction,

$$ZnSO_4 + H_2O \longrightarrow Zn + H_2SO_4 + 1/2O_2 \qquad (2)$$

Acid generated during electrolysis is utilised for further leaching. Zinc metal in the form of sheets is stripped off manually from aluminium cathodes after 24 hours and melted in crucible furnace and cast as zinc ingots. The generalised flow-sheet for the process is given in Figure 1.

Solvent Extraction and Electrowinning (SX - EW) Process

This is an integrated process with electrowinning approach. All processing steps other than purification step are similar to the process described above. In case of high copper containing material particularly brass ash, removal of copper by zinc dust cementation method is not attractive. Therefore copper is removed by solvent extraction to produce copper metal as shown in Figure 2. The solvent extraction process is carried out in a series of mixer-settlers which consist of three sub sections, namely (i) extraction, (ii) scrubbing and (iii) stripping.

After leaching and iron removal the clarified filtrate containing zinc and copper, comes in counter current contact with a suitable solvent in the extraction stage. Zinc is extracted selectively by the organic phase along with traces of copper. Copper is removed from the loaded organic phase in the scrubbing stage. The organic medium after scrubbing is brought into counter current contact with spent electrolyte in the stripping section. The resulting zinc electrolyte goes to cell house for zinc electro-winning and the barren organic is re-circulated to the extraction stage.

Similarly for copper circuit, a parallel stream of clarified feed is taken for extraction with the help of another organic, which selectively picks up only copper and the depleted aqueous phase is sent back to leaching as raffinate. Stripping of copper from loaded organic phase and electrowinning are more or less same as that of zinc.

Crystallisation Process

Zinc ash fines are separated from metallic granules in pulverisers fitted with cyclone separator and bag filters and then processed for zinc sulphate manufacture. The fine ash containing mostly zinc oxide is treated with sulphuric acid to bring zinc values to sulphate form and filtered to get clear zinc sulphate solution as shown in Figure 1. The solution is evaporated in a crystalliser and crystallised by cooling through chilled water. Pure crystal of zinc sulphate are separated from mother liquor by centrifuging and are dried, packed and sold in the market for micronutrients and other applications.

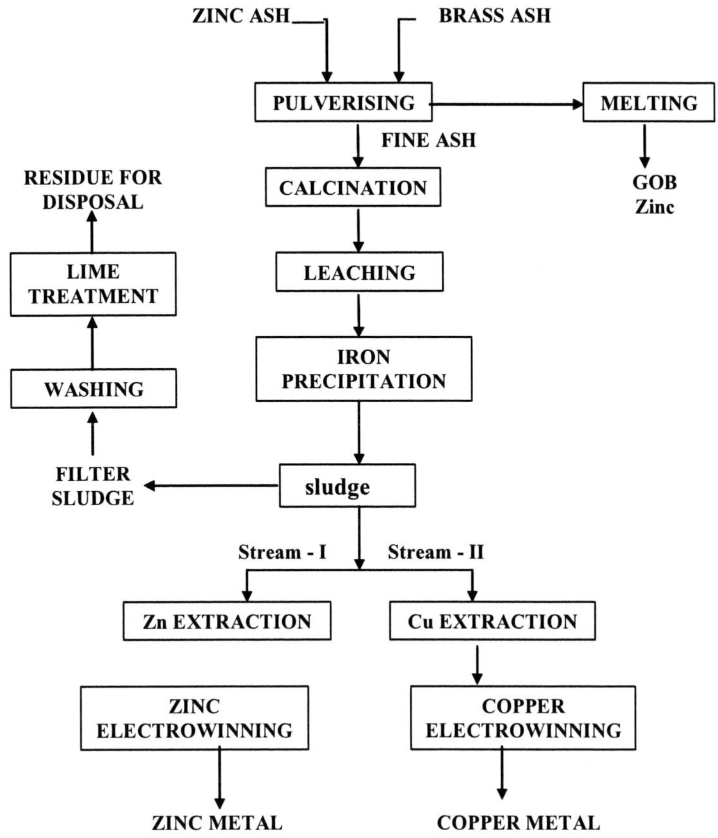

Figure 2. Block diagram for solvent extraction and electrowinning process to treat copper rich zinc secondaries

Newer Technologies for Zinc Recycling

Several new hydrometallurgical processes [7] have been developed which are considered to be energy efficient and eco-friendly. These are:

ZINCEX Process

This hydrometallurgical process resembles conventional zinc electro winning and is readily integrated with such a plant. EAF dust is leached with sulphuric acid to solubilize zinc and cadmium oxide and halides. The impure zinc sulphate solution is treated by solvent extraction to remove zinc and cadmium and transfer them to purified, halide free, concentrated zinc sulphate solution. Cadmium is removed from the solution by cementation, which is sold. High purity zinc metal is electrowon from the purified solution and sulphuric acid generated is recycled to the process. The zinc ferrite containing leach residue is further treated to recover lead and is finally suitable for disposal [25-28].

AUSMELT Process

Zinc is recovered from electric arc furnace (EAF) dust by bath smelting technology. Here oxygen, flux and coal are injected via a cooled lance into liquid slag in one furnace to melt the EAF dust and in the second furnace to reduce the zinc, lead and cadmium oxides. The fumed zinc, lead and cadmium are reoxidised above the bath and collected in a bag filter. Non-hazardous iron rich slag is prepared for sale or disposal [29]

EZINEX Process

This hydrometallurgical process is based on the leaching of the secondaries with ammonium chloride to solubilize zinc oxide, lead oxide and cadmium oxide. The leach solution is filtered, followed by purification of lead and cadmium by cementation on zinc dust and electrowon to produce high grade or hot dip grade zinc metal. The spent electrolyte is recycled to leaching. The iron rich zinc-ferrite containing leach residue is dried, palletized with coal, and recycled back to EAF, thus no byproducts are produced for disposal, Mixed KCl-NaCl salt is crystallized for sale as flux. [30-31].

WAELZ KILN Process

In the two stage Waelz kiln process the raw dust is fed to the first kiln, to separate Zn, Pb, Cd and chloride from the non-hazardous partially metallized iron. The dust from the first kiln is retreated in the second kiln to produce impure ZnO for Zn smelter feed and lead-cadmium chloride for processing to separate lead and cadmium.

The single stage Waelz kiln process is identical to the first stage of the two stage process. The ZnO is generally treated in the Imperial Smelting Process to produce zinc and lead metal or converted to zinc-based chemicals and fertilizer additives [7].

CENIM-LNETI Process

This is based on ammonium chloride leaching of secondaries, followed by solvent extraction and electrowinning / precipitation of zinc. This technique uses integrated processes to prepare various other products and has flexibility to treat different raw materials of variable compositions. The process can be adopted in small and medium scale industries [32].

CARBOFER Process

Recently a British service group to steel producers, gave an account of developments in the recycling of steel by-products by this new process known as Carbofer process. This process is used to recycle oily millscale at EAF operations. When post-consumer scrap is recycled, dust collected from furnaces contains significant amounts of zinc (5 – 20%), which can be recovered by the Carbofer process. A series of developments at a UK site has discovered that by including EAF dust in the blend, the zinc content of the dust is even higher, making it suitable for reprocessing by secondary zinc operations [33].

LEAD INDUSTRIES

India has two major primary lead producers namely Hindustan Zinc Ltd (HZL) with Visakapatnam plant producing 22,000 tons per year, Chanderia plant in Rajasthan with an annual capacity of 35,000 tons, and HZL Tundoo plant in Bihar producing 8,000 tons per year with a total of 65,000 tons per year [1] by all the three plants. India's largest refinary being setup by Sterlite-managed by Hindustan zinc group with an annual capacity of 50,000 tonnes inside the premises of HZL's Chanderia Lead Zinc smelter with the help of Australia based Ausmelt technology [34]. The Indian Lead Limited (ILL) with a capacity of 24,000 tons per year which is gradually shifting to the concentrate route due to the restricted availability of lead scrap. There are about seven medium size secondary lead plants (total capacity 55,000 tonnes/year), 40 small operations (combined capacity of 15,000 tonnes/year) and more than 250 tiny/backyard plants (estimated combined capacity 25,000 tonnes/year). The main source of lead in these plants is scrap batteries. The overall contribution of the secondary lead industry in the country is almost same as that of primary producers. Around 52,000 tons of secondary lead is produced from the organised sectors but only 30,000 tons are available in the open market as the remaining part goes to captive consumption. Lead production from unorganised sectors (backyard smelters) is about 15,000 tons per year. Thus by the end of ninth plan (1997-98 to 2001-02) the demand and supply gap was about 41,700 tons which would further increase to 114,500 tons and 192,200 tons by the end of tenth (2002-03 to 2006-07) and eleventh plan (2007-08 to 2011-12) [1] respectively, if the supply from the backyard smelters (bhatties) is not taken into account. Table 4 depicts the demand and supply details of lead in India[61]. The demand of lead, which cannot be met by primary production, can be compensated to some extent by secondary production.

Table 4. Lead-Demand-supply Gaps

Year	Demand (Tonnes) with 6% Growth	Supply	(Tommes)	Sec Lead	Total	Gap+ Surplus- Deficit(Tonnes)
		HZL	ILL			
10th Plan						
2002-03	137700	38700	21600	45000	105300	-32400
2003-04	146000	70200	21600	45000	136800	-9200
2004-05	154700	70200	21600	50000	141800	-12900
2005-06	164000	70200	21600	50000	141800	-22200
2006-07	173800	70200	21600	50000	141800	-32000
11th Plan						
2007-08	184300	70200	21600	6000	151800	-32500
2008-09	196100	70200	21600	6000	151800	-44300
2009-10	208600	70200	21600	6000	151800	-56800
2010-11	222000	70200	21600	6000	151800	-70200
2011-12	236200	70200	21600	6000	151800	-84400
12th Plan						
2016-17	316000	70200	21600	7000	151800	-164200

Lead is a wide spread constituent in the earth crust. It is virtually indestructible and has always been present in soils, water and air. It is extensively used by mankind and has been dispersed continuously over a long period of time into the environment. There are many ways by which lead can enter the environment. Lead acid batteries, their handling and reprocessing in an unorganised manner is one of the major source of environmental pollution. Today, the lead acid battery scrap both in developed and developing countries are the single largest source of lead metal in meeting the demand. The reason is lower capital investment and energy requirement per tonne of metal production. In India there are only few industries in the organised sector who are having pollution control facilities, however, in the unorganised the situation is totally different. Most of the waste lead acid batteries are reprocessed by the unorganised sectors causing problem to the workers and the nearby locality. This is mainly due to improper technology, cash crunch and lack of monitoring and implementation of pollution norms. For ecological awareness and environmental consciousness, it is necessary to control pollution particularly in the unorganised lead production units.

Currently over 70% of the lead consumed by automobile industry world-wide is fed by the secondary lead smelting plants, mainly derived from the used acid batteries. Besides battery scrap there are a wide variety of damaged/old scrap, obsolete or worn-out products including bearing metals, sheets and pipes, cable sheathing and solder as well as dust and drosses containing lead. All these fall under the category of restricted materials and hence must be handled with utmost care owing to the toxic effects of lead and dangers of the sulphuric acid. Non-separation of the scrap batteries/compounds and smelting the whole battery with its plastic body produces poisonous dioxine, which is detrimental and hazardous to the public health causing damage to the environment and the people alike.

Waste Treatment/ Disposal Practices in Lead Industries

The details of waste generated, their quantity, composition and disposal practices of some of the major lead plants are incorporated in Table 5. Dumping or stockpiling of lead blast furnace slag, ISF slag and flue dust is mostly followed in primary lead units. Though the slags are vitreous in nature this mode of disposal is now questionable on environmental and economical grounds. The by-products of these units, like enriched silver crust/prest crust and copper dross in HZL plants, are mostly processed for the recovery of silver and lead metals. Matte and speiss generated are stored or sold for copper recovery. Antimony dross of HZL Chanderia is stored or sold to battery makers. The ferrosilicate slag containing >2% lead generated in Indian Lead Limited is recycled to extract lead and discarded only after lead content is brought down to <2%. Presently the slag is stored inside the factory premises for want of suitable land site allocation by the respective State Pollution Control Board (SPCB).

Since lead is a very toxic material, disposal of the solid wastes in the secured landfill is absolutely necessary. However, major problem is with the secondary lead smelting units and most of them do not follow the proper processing technologies and are handling the spent lead acid battery scraps without resorting to the pollution control norms with respect to the SPM level, SO_2, CO and lead fumes. During the processing of spent lead acid battery in backyard units, operations like breaking, crushing, screening, dry mixing etc. generate airborne lead dust which directly or indirectly enter into the human system and the surroundings of the working area. Lead absorption into the human body in significant quantity causes lead poisoning. Several of these backyard smelters simply recover 60-70% of the metal using home made technologies because of the low melting point of the metal and simply throw the slag containing very high lead into the environment causing greater damage to nearby locality. The worst sufferers in the unorganised backyard smelters are the workers involved in these units and the environment affecting the local population.

Table 5. Details of metallic waste in primary lead industry

Sl.No	Name of the Industry	Waste Generated: Quantity & Composition (%)	Waste: Reprocessing/ Dumping
1	Zinc lead smelter, HZL,Visakhapatnam[62] (Presently shut down)	**-BF Slag**: 15,000 t/y Pb: 1.5-2.0, SiO_2: 20-22, FeO: 36-38, MnO_2: 3.0, CaO: 14, MgO: 2.0, Al_2O_3: 6-7, Zn: 5-7, S: 2.0 **-Ordinary Cu dross** Pb: 83.0, Ag: 0.09, Cu: 2.0, Zn: 1.0, Sb: 0.3, Bi: traces **-Antimony dross** Pb: 70.0, Ag: 0.05, Cu: 0.3, Zn: 1.0, Sb: 2.0, Bi: traces **-Final refinery dross** Pb: 72.0, Ag: 0.03, Cu: 0.2, Zn: 5.2, Sb: 0.5 **-Enriched liquated silver crust** Pb: 78.0, Ag: 5.8, Cu: 0.7, Zn: 14.0, Sb: 0.1	-Part of the slag is recycled & rest is dumped
2	Tundoo lead smelter, HZL, Bihar[63]	**-BF Slag**: 18, 000 t/y Pb: 1.5-2, FeO: 32-35, ZnO: 9-12, MgO: 4-4.5, Insoluble: 23-24, CaO:14-16, Al_2O_3: 9-10, Ag: 0.001-0.002 **-Copper & ordinary drosses:** 300 t/y Cu: 2-4.5, Pb: 80, As: 10-11 **-Prest Crust**: 375 t/y Ag: 4-5, Zn 15-18, Rest: Pb	-Part of it is recycled in the sintered section. Rest is dumped -For silver lead recovery -Goes For Ag recovery
3	HZL Chanderia, Rajasthan[11,12]	**-Dezincing & antimony drosses:**1.5 t/y **-Litharge:** 480 t/y **-ISP Slag**: 42,000 t/y Zn: 6-9, Pb: 0.6-0.7, Cu: 0.19, Sb: .01, Ag: 19 g/t, Au: 0.04 g/t, FeO: 33	- - -Dumped and commercial disposal
4	Indian Lead Ltd., Thane[4]	**-Matte & Speiss:** 1, 300 t/y Pb: 17.3,Cu: 44.6, Sb: 1.2, Ag: 19.50 g/MT, Au: 1.9 g/t **-Antimony drosses**: 1, 520 t/y Pb: 51.5, Cu: 0.9, Sb: 7.0 **-Silver crust (enriched):** 370 t/y, Pb: 24, Cu: 2.4, Ag: 15 g/t, Au:0.012 **Ferrosilicate slag**: 3,000 t/y FeO:40-50, CaO: 10-20, SiO_2: 35-45, Pb: <2.0	-Commercial disposal for Cu recovery or stored -Stored or sold to battery makers -Reprocessed for Ag recovery -Slag containing lead >2% is recycled & discarded only when lead content is <2%.

Efforts need to be made to improve health protection of the workers and to reduce the quantity of lead entering to the environment, which can be met through

- Efficient collection of the acid battery scrap.
- Strict control over unorganised sectors.
- Modified working practice and adoption of eco-friendly process technologies.

Secondary Lead Processing

Lead is a versatile and strategic metal with many vital applications. Automobile sector is the largest consumer of lead because of its use in the production of lead-acid batteries. The total market for lead in the world is actually quite large, it's about seven million tonnes. A large component of that comes from recycling electrical battery units and about 40 per cent of the world's lead comes from primary sources [35]. Lead scrap in the main sources of secondary lead as more than two-third of the total lead produced is consumed battery sector. The battery manufacturers consume about two-third of the total lead produced. Lead is also used in the sheathing of a variety of cables e.g., fibre optics cables, as a shielding material for radioactive radiations because of it specific properties like low melting point, ductility and high density. Lead dampers have been developed for earthquake protection. Other uses of lead are protection coatings, paints, solders, and gasoline additives. However use of lead compounds as gasoline additives is decreasing in third world countries too because of the concept of unleaded automobile fuel. As already mentioned primary lead resources are depleting and environmental awareness is increasing, causing the concept of reuse/recycling to become mandatory which suits to lead recovery along with the recovery of other precious metals and non-ferrous metals. In fact primary resources are quite limited and major amount of lead production is through secondary processes. The various processing steps [36] of scrap battery are shown in Figure 3. The metallic lead of grids can be readily melted at relatively low temperatures and cast into ingots of secondary lead bullion. Whereas, the paste requires a more complex recovery operation.

Almost all the secondary plants use the pyro-metallurgical smelting process. The newer secondary plants generally have larger smelting capacities, which are more efficient and environmentally safer. Globally over 60% of total lead is recycled. Secondary lead production in other countries is shown in Table 6 [61]. Battery scrap from automobile sector accounts for 80% of old scrap recycled as secondary lead raw material. A standard lead acid battery for starting, lighting, and ignition of vehicles has the following average composition by weight[37]:

A	Lead metal	34%
B	Lead oxide paste	39%
C	Polypropylene	5 - 6%
D	Electrolyte (free sulphuric acid)	11-12%
E	Others (ebonite, PVC, paper etc	8 -10%

Table 6. Secondary lead production-world scenario

Country	Secondary lead production-share(%)
USA	35.7
Germany	7.1
Japan	6.0
UK	5.9
Italy	5.3
France	4.5
Canada	4.1
Spain	3.9
Belgium	3.5
Others	24.1

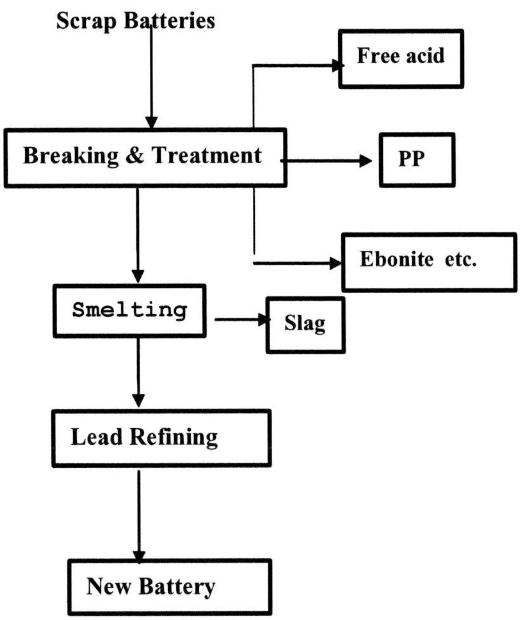

Figure 3. Flow sheet for recycling of lead acid battery adopted by Unorganized sector. (pp: polypropylene)

The pyrometallurgical smelting process employed at almost all secondary plants is essentially same as that used in the primary smelter, there are key difference in feed materials and the type of smelting furnaces in use. Choice of secondary smelting process is independent of the battery breaking procedure used, although there are difference in the type and form of feed preferred by various furnaces. The newer secondary plants generally have larger smelting units, which are more efficient and environmentally safer. The various processing steps [38] of scrap battery are:

- Separation of all the constituents namely acid, polypropylene, separator and ebonite, metallic and non-metallic lead fractions
- Smelting of metallic and non-metallic fraction
- Refining of lead along with production of lead and lead compounds of desired composition

Battery breaking and the feed preparation is the first operation in most plants. Through a sequence of crushing, screening, washing, heavy media separation and floatation process, segregated streams of battery grid metal, battery paste, polypropylene, rubber and waste acids are produced. The metallic lead of grids can be readily melted at relatively low temperatures and cast in to ingots of secondary lead bullion. Whereas the paste requires a more complex recovery operation as follows:

- Desulphurisation, i.e. removal of sulphur from the lead sulphate by conversion to lead carbonate (the most common desulphurisation process used is sodium carbonate mixed with the oxide and sulphate paste).
- Reduction (lead carbonate and unreacted lead oxide into metallic secondary lead bullion, i.e. unrefined and impure lead) and
- Refining of secondary lead bullion into soft lead, lead calcium alloy and/or antimonial lead

It is most unlikely that backyard recyclers will have a furnace capable of recovering the lead from the paste and the sludge of scrap batteries. They use an open kiln or a pot called *bhatties* for smelting. The metallic lead in the battery grids is melted and cast into ingots (Figure 3). The recovery rates are thus hardly higher than 60% of the available lead in the scrap battery and consequently there are serious pollution problems.

The pollution problems in lead industries are of great concern as they persist despite of modest provisions for the control of lead dust and fumes and for the protection of workers in the occupational environment. Keeping in mind the hazardous nature of lead, it is necessary to control pollution particularly in the unorganised secondary lead production units. Some of the drawbacks found in India are summarised as follows:

- During opening and breaking, the sulphuric acid is allowed to wash down to the street drain and percolate in to the ground and contaminate the ground water. Acids tipped in to the streams and river will lower the pH of the water and adversely affect the local ecosystem.
- The procedure to open the spent batteries and change the plates or remove certain cells causes the risk of acid splashes to the skin and eyes to the person undertaking the task without proper safety measures.
- The bags are opened manually and the contents are unloaded at the charge preparation area adjacent to the furnace. Cloudy dust generation during this activity contributes to the elevated lead levels in the atmosphere.
- Smelters in the unorganised sector are not equipped with proper facilities of recovering the lead from the paste. The residue after recovery of lead contains high lead and causes serious problem of pollution in case of disposal.
- During the smelting process sulphur in the charge comes out as sulphur dioxide and emit large amount of dust and solid particulate matter (SPM) containing high lead causes health hazards.
- During handling and processing of battery scrap such as breaking, charge preparation, smelting, refining and packing, very little precautions are taken by the workers.

Operations like preparation of alloys, oxide, powders, paste and dry mixing of oxides, filing, assembling etc. leads to air borne dust pollution. In the dry process, lead concentrations in the air ranges from 0.13 – 3.1 mg/m^3. Emission factor for typical uncontrolled emission is 8 kg of lead per 1000 number of batteries [39]. The 80% of which can be checked by simple measures in and around the pollution areas. Emitted lead is directly or indirectly entered into the body of the workers and others surrounding the working area. Uptake of lead in to the body through inhalation is more. About 48-64% of the inhaled lead is deposited in the lungs. The gastrointestinal absorption is only 15%.

Like anywhere else, India has sound technology for the recovery of lead via the recycling route in the organised sectors. Backyard recycling of lead from acid batteries is the most environmentally unacceptable practice and this has increased dramatically over the years in spite of the provisions of the Central Pollution Control Board. Also such materials are being imported from several countries even though India is a ratified signatory of Basel convention.

In the recent past, a number of ecofriendly pyrometallurgical processes like, ISAMELT, AUSMELT, KIVCET, QSL, KALDO process etc. that have been developed and commercialised to process ores and concentrates are also suitable for treating secondary lead particularly the battery scrap. MRU process followed in Germany is quite eco-friendly in terms of energy consumption and lower Pb and SO_2 emissions (Table 7). Table 8 shows a comparison between these primary lead smelting operations with regards to their ability to process secondary lead too [7]. QSL plants operating in Germany, Canada and Korea are using 30-35% lead secondaries in the feed [40] These technologies meet the norms of western countries and could also be applied in countries like India.

Table 7. Energy requirement Pb and SO_2 emission from different smelter technologies

Item	Primary lead production		Secondary lead production
	Sinter shaft furnace	QSL furnace with energy recovery	MRU technology
Energy [MJ/Mt] of lead	10,000	7,000	6,000
SO_2 emission [kg/Mt of lead)	70	4	1
Lead emission-air[g/Mt of lead]	140	40	7

Table 8. New primary smelting processes and their ability to accept lead secondaries as feed

Process	Type of Smelting	Mode of Operation	Appetite for Secondaries	Comment
QSL	Bath	Continuous	Significant	Several plants are in operation
Outokumpu Oy	Flash	Continuous	Little	Separate smelting/slag reduction furnaces
Kaldo (Top blown rotary converter)	Flash/Bath	Batch	Significant	Rotates and tilts, mechanically complex
Isasmelt	Flash/Bath	Batch	Significant	Lacks flexibility in feed
Kivcet	Flash	Continuous	Little	Kivcet-CS can process 100% oxidised feed

Like the primary lead smelting operation, secondary plants also use pyrometallurgical processes with a difference in feed material and the type of smelters used. Table 9 shows the main feature of pyrometallurgical processes applied in the secondary lead production [7, 39-41]. Several new hydrometallurgical process for the recovery of high pure lead from spent lead acid battery have been developed in last two decades which has several advantages such as no dust and fumes generation, flexibility in scale of operation, location, environmental friendliness etc over thermal processing should be encouraged. Most of these processes are either on laboratory scale or tested on pilot scale of operation. The following text describes the processes in brief.

Table 9. Main features of secondary Lead Smelting Processes

Process	Smelting /Mode	Features/Current Status
Blast Furnace	Shaft/Continuous	Can be used for a range of secondary feed like scrap lead, slags, drosses and flue dust. Widely used world wide Inherently non clean technology producing high silica mattes and slags.
Reverberatory Furnace	Bath/Continuous or Batch	Used primarily for high oxidic and antimony containing feed Widely used worldwide Environmental problems.
Short Rotary Furnace	Bath/Batch	Used primarily for high metallic feed Widely used world wide Produces a reactive soda slag which can cause disposal problems.
Kaldo Furnace	Flash or Bath/Batch	Used for smelting residues and scrap including unbroken batteries In operation in Sweden on secondary feed.
Electrothermic Furnace	Bath/Batch	Uses metallic and oxidic feed Russian companies offer the technology Costly equipments Status not known.

RSR Process

In this process the $PbSO_4$ portion of the sludge is treated with ammonium/alkali carbonate to produce lead carbonate and ammonium sulphate.

$$PbSO_4 + (NH_4)_2CO_3 \longrightarrow PbCO_3 + (NH_4)_2SO_4 \quad (3)$$

Reduction of PbO_2 is done in two ways:

- By addition of SO_2 to an alkali/ammonium carbonate producing alkali sulphite/bisulphite which reacts with PbO_2 to produce $PbSO_4$. Sulphite/bisulphite are oxidised to sulphates and lead is precipitated as $PbCO_3$ or $PbCO_3.Pb(OH)_2$.
- By heating the battery sludge to 290 °C in presence of some amount of organic.

The desulphurised sludge is leached in H_2SiF_6 and HBF_4 and lead is electro won. About 500 ppm of arsenic is added to the electrolyte in order to prevent PbO_2 deposition on the anode. The leach solution may contain 70-200 g/L and at least 50 g/L of free acid.

Insoluble PbO_2, part of the sludge, can be reduced to soluble form either by addition of SO_2 to alkali/ammonium carbonate producing alkali sulphite / bisulphite which reacts with PbO_2 to produce $PbSO_4$. Sulphite / bisulphite is oxidised to sulphates and lead is precipitated as $PbCO_3$ or $PbCO_3$. $Pb(OH)_2$, or by heating the battery sludge to 290 °C in presence of some amount of organics. This treated sludge is leached in H_2SiF_6 and HBF_4, followed by electrolysis [42].

Bureau of Mines (USBM) Process

Like RSR process the sludge (containing $PbSO_4$ and PbO_2) is pre-treated with ammonium carbonate to desulphurise $PbSO_4$. Here the reduction of PbO_2 to PbO is done by addition of metallic lead powder during leaching with H_2SiF_6. The drawback of this process is the requirement/recycling of large amount of powder lead, as the PbO_2 content in the sludge is about 35-40%. This process utilises PbO_2 coated titanium anodes and lead cathodes and about 1-2 g/L of phosphorous (using salts of phosphoric acid like, $Na_3PO_4.12\ H_2O$, $(NH_4)_2HPO_4$, $CaH_4(PO_4)_2$ to prevent PbO_2 deposition on the cathode [43]. The lead content in the electrolyte may be as high as 150 g/L and depleted lead level is above 25-35 g/L.

Engitech Process

In this process the battery is broken to separate various fractions and desulfurize the sludge using Na_2CO_3 and NaOH to produce Na_2SO_4 which is crystallised to produce the crystals of Na_2SO_4 [44] The desulphurised sludge containing Pb, $PbCO_3$, PbO, PbO_2 and $Pb(OH)_2$ is leached with HBF_4 (Eq. 2). The process also uses H_2O_2 to dissolve PbO_2 (Eq. 3).

$$PbO_2 + Pb + 4HBF_4 \Longleftrightarrow 2Pb(BF_4)_2 + 2H_2O \quad (4)$$

$$PbO_2 + H_2O_2 + 2HBF_4 \Longleftrightarrow Pb(BF_4)_2 + 2H_2O + O_2 \quad (5)$$

To avoid PbO_2 diposition at anode, the Engitech process [45] utilises a specially designed composite anode, that operates at an extremely high anode current density e.g(320 A/m^2). Lead is electrowon at this high current density.

Ginata Process

In this process lead is extracted from all the ectrodes, and electrical connections of spent lead-acid electric storage batteries. The spent lead-acid battery is recharged to reduce the $PbSO_4$ removing sulphuric acid from the recharged battery. The battery is immersed in an electrolyte (like Fluoborates, Fluosilicates, Dithionates, Perchlorates, Cyanides, Nitrates, Oxalates, and Pyrophosphates) in which lead from all the electrodes and electrical connections of a spent battery will form a complex with the electrolyte. Since each immersed battery is an electrochemical system composed of a series of electronically connected positive and negative couples of electrodes all immersed in the same electrolyte, in which the negative electrodes naturally dissolve anodically to release lead ions into the electrolyte thereby reducing Pb^{4+} to Pb^{2+}, which dissolves in the electrolyte as lead complex [46]

The resulting electrolyte containing lead is electrowon in presence of about 200 ppm cobalt as additives eliminating the need for PbO_2 coated or special alloy anode. Cobalt enhances the evolution of oxygen and thus minimises the degradation of graphite anode.

The electrowinning process is operated under periodic reverse current to oxidise antimony in the electrolyte to pentavalent state, which is precipitated from the electrolyte above the concentration of 3,000 ppm. High pure lead cathode is produced in the process.

AAS Process

This process utilises ammoniacal ammonium sulphate (AAS) solution for the leaching of the battery scrap. The whole crushed batteries are fed to an upward moving leaching column, where the solution floats the case and separators out of the column, suspends and dissolves the lead compounds of the paste, separating the clean metallic lead fraction. The PbO and $PbSO_4$ dissolve in AAS solution and undissolved PbO_2 is separated and converted to $PbSO_4$, using about 50% H_2SO_4 at high temperature. The resulting $PbSO_4$ is fed back to the AAS leaching process. The lead is electrowon to get sponge lead. This process utilises ammonia and water as feed and recovers $(NH_4)_2SO_4$ as byproduct through a bleed stream [48-49].

$PbSO_4$ Slurry Process

This process involves direct electrowinning of lead from the battery paste. $PbSO_4$ and PbO from the battery paste are introduced in the cathode compartment by an ion selective membrane. The electrolyte of the cathode compartment consists of the slurry of $PbSO_4$, PbO, H_2SO_4, Na_2SO_4, NaOH along with a complexing reagent like ethylene diamine tera-acetic acid (EDTA), acetic acid, oxalic acid etc. The cathode is a fluidized bed of particles connected to a current collector. During electrowining process, the slurry particles contact the cathode and are reduced to metallic lead. The sulphate ions diffuse through the membrane forming H_2SO_4. The PbO_2 in the slurry is contacted with SO_2 or Na_2SO_3 to produce $PbSO_4$ which is recirculated in the cathode chamber. In this process the battery paste is converted to metallic lead and suphuric acid [50].

PLACID Process

This process involves direct chloride leaching rather than disulphurisation and leaching with HBF_4 or H_2SiF_6. The chemical equations of chloride leaching of battery paste are:

$$PbO + 2HCl \longrightarrow PbCl_2 + H_2O \quad (6)$$

$$Pb + PbO_2 + 4HCl \longrightarrow 2PbCl_2 + 2H_2O \quad (7)$$

$$PbSO_4 + 2NaCl \longrightarrow PbCl_2 + Na_2SO_4 \quad (8)$$

The soluble impurities are precipitated with lead powder as:

$$Pb + MeCl_2 \longrightarrow PbCl_2 + Me \quad (9)$$

The $PbCl_2$ is dissolved in HCl and fed into the cathode compartment of a diaphragm cell for lead electrowining.

Though the PLACID process [51] and other chloride processes are attractive from environmental point of view but have not found favour because lead is recovered as a sponge and large amount of lead powder is required to reduce PbO_2. A comparative view of this process with other processes is given in Table 10. A conceptual block diagram for Placid process is shown in Figure 4.

Table 10. Comparision of Placid Process with other electrowinning processes

Process	Electrolyte	Additive	Pb [g/L]	Purification	Cell type	Current [A/m^2]
RSR	Fluosilicic or fluoboric acid	>500ppm As + boric acid	70-200	None	Simple	100-200
USBM	Fluosilicic acid	3-6g/L H_3PO_4 + 0.05g/L Bone Glue + 4g/L Lignin sulphonate	25-150	None	Simple	150-250
Engitec	Fluoboric acid	Glues	50-100	None	Composite Anode	320
AAS	AAS	None	-	None	Cascaded	-
Placid	HCl Brine solution	none	25	cementation	Ion Selective Membrane	1200

AAS: Ammoniacal Ammonium Sulphate

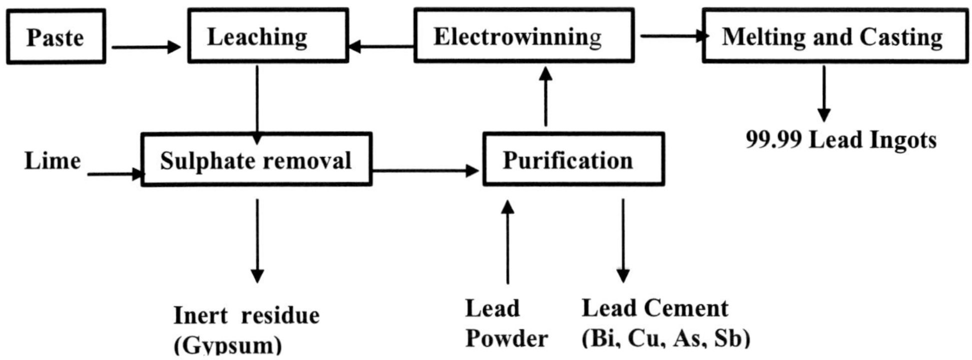

Figure 4. Placid Process (Conceptual block diagram)

PLINT Process

This process is almost similar to that of PLACID process [52, 53]. The only difference between this and the PLACID process is in the substitution of a precipitation step for electrowinning. In the subsequent kettle the lead hydroxide product is first decomposed and then reacted with hot coal to obtain pure lead. All of that takes place at a temperature not higher than that required for casting and alloying. The percentage recovery and the purity of the lead produced is same as in case of PLACID process, as the leaching and purification process remains unchanged. A conceptual block diagram is depicted in Figure 5.

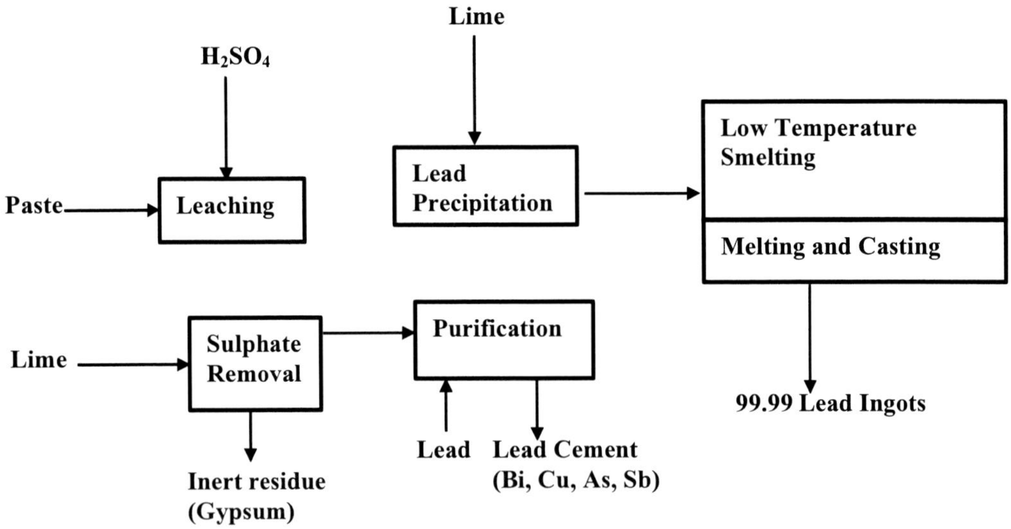

Figure 5. PLINT process (Conceptual Block Diagram)

Lead Pollution and Permissible Limits

Lead if absorbed into the human body in significant quantity, can pose a considerable threat to health. Individuals are at risk whether exposed to a single, large emission of lead, or facing prolonged low-level exposure. Lead poisoning occurs mainly in three organ systems[54].

- Red blood cells and their precursors
- Central and perial nervous systems
- Renal system

Acute lead poisoning in human causes severe damage to the kidneys, liver, brain and the central nervous system. It has also been associated with carcinogenesis, reduced fertility, miscarriages and spermatotoxicity. It causes anaemia, colic and mental disorders viz. defective memory, mental dullness, fatigue, nervousness and anxiety. Lead exposure has also been found to cause hypertension, which causes thousands of deaths.

The pollution control enforcement authorities are well aware of the ill effects of various means. The pollution control laws/rules in India have the same teeth as in the developed countries. However, its implementation is the main draw back of the system. For examples the norms of emissions from the stack as well as in the ambient air are as stringent as in any developed countries. A comparative statement of the norms is given in Table 11. Emission standard for SPM recommended by Maharastra Government in 1985 is 100 mg/m^3. The organised smelters, being easily identifiable, come under the close scrutiny of the authorities and unless they meet the stipulated standard, they are not allowed to continue with their operation. Such units have started taking some corrective measures.

Table 11. Emission control standards for lead plants

Country	Lead in work area, mg/NCM	Lead in blood, mg/100mL	Lead in emissions mg/NCM	Lead in effluents mg/L	Lead in atmosphere mg/NCM
Australia	0.15	70 * 30 †	10	0.05	1.5
Austria	0.10 * 0.02 †	100 *	10	0.05	-
Canada	0.15	70	29	15	-
Denmark	0.10	50	#	0.1-0.5	2.0
EEC	0.15	70	-	-	2.0
Germany	0.10	70 * 30 †	5	0.2	2.0
India	0.15	80	10	0.1	1.5
UK	0.15	70 * 40 †	10	variable	2.0
USA	0.1	50	variable	variable	1.5

* men.
† women.
# limit set for individual plants.

Indigenous generation secondaries are required to be recycled, otherwise the disposal/dispersal of the same will cause greater harm to environment. A range of restrictions on the collection, handling, transport and processing of spent batteries are require monitoring of their movements both domestically and internationally. Generally it is the battery components that cause the difficulties not the lead as such.

Table 12. Contaminants determined in TCLP test.

Sl. No.	Contaminants	EPA Hazardous Waste No.	Maximum Concentration (mg/L)
1	Arsenic	D004	5.0
2	Barium	D005	100.0
3	Cadmium	D006	1.0
4	Chromium	D007	5.0
5	Lead	D008	5.0
6	Mercury	D009	0.2
7	Selenium	D010	1.0
8	Silver	D011	5.0

Extraction procedure for toxicity assessment is designed to identify wastes that are likely to attain hazardous concentrations of the particular toxic constituents into ground water as a result of improper management such as dumping. In this procedure the wastes are leached in a manner to simulate the leaching action that occurs in the landfills. The extract is analysed to determine if it possesses certain toxic contaminants. If the concentration exceeds the regulatory levels then the waste is classified as hazardous. Some of the regulatory levels [3] of the contaminants in the leached extract are given in Table 12 as per guidelines of Environment protection Agency (EPA), USA which is accepted in India as well. It is therefore essential to test all the solid wastes from nonferrous metal industries in general and copper, zinc and lead primary and secondary units in particular, that are dumped without following proper procedure or used for land filling.

The lack of informations in open literature on TCLP test results, clearly demonstrated the level of negligence prevailing in these industries on solid metallic waste management. The solid wastes produced from the liquid effluents are no longer the safer options unless these wastes are handled and managed/processed in eco-friendly manners adopting specified procedures laid down by the pollution enforcement agencies. NML, Jamshedpur is presently involved in the characterisation of solid wastes that are generated in Indian nonferrous metal industries to assess the environmental degradation.

Basel Ban and its Implication

Approximately 5 million tonnes of wastes are generated annually without a single site identified for disposal. In addition, India imports close to 2 million tonnes of lead and zinc waste. The hazardous waste management and handling (HWM & H) clearly prohibit in trade of wastes for dumping and impose that only recyclable waste could be permitted after clearance from the Ministry of Environment and Forest (MoEF) and State Pollution Control Board (SPCB). But in subsequent years, the HWM&H regulations were not reflected in the import procedure and hundreds of tonnes of waste were freely imported under open general licence category (OGL). Some wastes imported were not suitable for recycling because the residues generated are toxic. Ban on import of some hazardous wastes has been imposed in order to prohibit and restrict further toxic residual waste generation. In view of the problems arising due to ambiguity in the identification or characterisation of some of the wastes being imported as raw material in the industrial process, there is a need to classify the wastes in terms of their hazardous contamination. The presence of hazardous constituents in the secondary materials invites caution while processing and disposing off the residues or wastes generated after processing/treating such materials. Between 1992 and 1998, India has been a major destination of metal wastes which actually were meant for recycling but often found its way into dumping sites. Indiscriminate imports of wastes in violation of Basel Convention and irregularities in reprocessing the wastes have had serious impacts on environment in various parts of the country.

Thus Basel convention was conceived as a means of preventing dumping of hazardous waste by the developed nations in countries, which do not have the facilities to handle them safely. India being a party to the Basel convention on the control and transboundry movement of hazardous wastes needs to regulate the import (transboundry movement of hazardous wastes) of such wastes under this framework. India signed the Convention on 15.3.1990, ratified it on 24.6.1992.

Since April 1995, only usable wastes were permitted for the imports under license for Directorate general of Foreign Trade (DGFT). In May 1997 Supreme Court of India banned import of all wastes for recycling and subsequently constituted an expert panel to study the complex issue of recycling of waste. In the mean while the pressure from industry and industrial bodies forced the Government to lift the ban on zinc ash and skimming, citing economic reasons. In December 1997, the Planning Commission recommended lifting the ban for trade on zinc ash and skimming, citing that domestic zinc production was inadequate for local needs [55]. Accordingly, HW Rules, 1989, were amended on 6.1.2000 ostensibly to improve the applicability and implementation aspects with regard to imports of hazardous waste, and to bring the Rules in line with the requirements of the Basel Convention. The provisions relating to import and export of hazardous waste for recycling have been expanded and streamlined. A time limit has been imposed for processing of applications for import. A fee for processing of import licenses has been prescribed.

As per the revised HW Rules, 1989/2000, the MoEF will grant permissions to importers/exporters and must satisfy itself that the importer has environmentally friendly/appropriate technology for reprocessing; that the importer has the capability to handle and reprocess hazardous wastes in an environmentally sound manner; and that the importer has adequate facilities for treatment and disposal of wastes generated. The CPCB is now responsible for ensuring implementation of the conditions of imports and for monitoring compliance with the conditions of import. The DGFT will now issue or refuse licenses for import of hazardous wastes with reference to whether the relevant permissions have been obtained by the importer from the MoEF.

Recent Development on Secondary Zinc and Lead Processing in India

Recently some R & D work for the recovery of zinc from galvaniser's ash and similar by-product compounds obtained as waste from various industries is carried out at various Council of Scientific and Industrial Research (CSIR) Laboratories in India. The optimum conditions were established on 500A scale, for producing 10 kg of zinc per day at Central Electrochemical Research Institute (CECRI), Karaikudi and a novel method (suspension electrolysis) has been developed for the recovery of zinc. The process has been commercialized. The zinc powder obtained is suitable for the production of hydrosulphite [1].

Regional Research Laboratory (RRL) Bhubaneshwar, developed an expertise for the processing of a variety of raw materials containing zinc like complex, dirty raw materials and secondary sources like waste/by-products from various primary zinc producers by hydrometallurgical routes which consists of pre-treatment, leaching, solid-liquid separation, solution purification solvent extraction and pure zinc metal recovery by electrolysis or recovery of a pure salt like zinc sulphate or zinc oxide. A method has been developed for the production of zinc and copper from spent catalyst from fertilizer industry at Kanpur by hydrometallurgical route [56]. Another process for the recovery of zinc from brass ash containing 30.6% zinc and 7.3% copper, and zinc ash containing 78.5% zinc, 2.9% chloride and other impurities like Cu, Fe, Ni, etc. to produce high pure cathodes of zinc is established [56]. Chloride was removed by roasting the zinc and brass ash at about 950 °C, followed by water washing and acid leaching with a leaching efficiency of 90-95%. The leach slurry is filtered and leach liquor is subjected to purification steps. In case of brass ash, copper is removed and recovered as copper powder first under controlled and optimized electrowinning conditions. The purified leach liquor is then sent for electrowinning to produce pure zinc metal. On the basis of the process know-how, M/s Panthnagar Fertilizer is producing 350 kg/day zinc and 50 kg/day copper. With the operating condition for zinc electrowinning as Zn 55-65 g/L, H_2SO_4 110-125 g/L, organic additive 20-30 mg/L, current density 350-400 A/m^2, cell voltage 3-3.2 V, current efficiency >90%, energy consumption 3,000 kWh/ton to produce a smooth, coherent, perfect sheet which was easily strippable.

A flow sheet has also been developed for zinc oxide production from zinc ash with zinc content of 60-80% along with other impurities like Fe, Cu and Pb for a private party in Kolkata. The unit operations involved are dissolution of zinc ash, purification of leach liquor, preparation of zinc hydroxide/carbonate mixture and finally zinc oxide by controlled heating of this hydroxide/carbonate. The purity of ZnO is comparable to international standard [56].

Recently the drum filter cake (DFC) generated from zinc industry(Composition SiO_2: 20-55%, Fe: 5-20%, Al: 5-25%, Pb: 1-5%, MnO_2: 1-10%) was used to prepare frit/glaze for ceramic industry[57] at RRL Bhopal. Silica present in DFC acts as glass former and lead as a flux for glass formation. Toxic lead gets complexed in the glass matrix to make the product non hazardous. The product was tested for its toxicity following TCLP test[57].

National Metallurgical Laboratory, Jamshedpur is actively involved in developing both pyrometallurgical as well as hydrometallurgical processes to recover zinc from zinc dross/ash and other residues. There are no efficient pyro/ hydrometallurgical processes available in India to treat the zinc dross and Effluent Treatment Plant (ETP) sludges without generating solid waste for disposal. Production of zinc oxide from ash by ammoniacal route developed earlier at NML has been found to be technically competitive to the sulphuric acid route [58]. A direct electrowinning process to recover zinc as zinc powder from an alkaline electrolytic bath is being developed on laboratory scale. The process requires less number of steps and is thus expected to be energy efficient. Zinc recovery from the dross by distillation process has been developed recently at NML with 90% overall recovery, from the dross produced in the galvanising plant of TATA STEEL.[24, 59-61].

From the environmental point of view, lead recovery from battery sludge via hydrometallurgical and electrowinning processes looks attractive. But none of the above-described processes have become operational and only a number of pilot plant trials have been conducted. Some processes developed are at laboratory scale only. A full-scale lead hydrometallurgical and electrowinning plant need to be operational in the coming century to build the confidence among the lead producer and to reduce environmental impacts. Use of cupola for lead recovery from battery scrap is also being developed in NML. R&D work related to the lead recovery from spent acid batteries is also being carried out at CECRI, Karaikudi. A novel direct electrochemical reduction technique has been developed for the production of lead powders from spent lead acid batteries [1].

Binani industries is setting up an integrated battery-recycling unit in Wada, Thane District of Maharashtra, which would be implemented in two phases. The first phase was planned be for the manufacture of 25,000MT of secondary lead while the second phase was for the manufacture of 125,000MT of Primary Lead with imported Lead Concentrates. The plant would be one of India's first modern lead recycling plants. ENGITEC Technologies of Italy will provide the technical know- how for the manufacture of lead. This technology is not only environment friendly but also has some of the best safety measure in the industry. The recycling facility will also yield sulphuric acid and polypropylene as by-products. Manufacturers of lead-acid batteries, pigments, lead oxide and submarine cables would be the main users of the lead.

In order to fix the limits of toxic elements in Zinc ash/skimming an effort was made at NML, based on the data available for head analysis and TCLP tests of several samples and recommended tentative threshold limits (Table 13) of metal components in zinc ash/skimming for import/export purposes. Based on the recommendation, the Ministry of Environment and Forest permitted import of zinc secondaries to a number of recycling units possessing necessary processing technology and waste handling and disposal expertise/facilities, in the overall interest of the nation.

Table 13. The tentative limits fixed for toxic elements in zinc ash for imports

Sl. No.	Components	Percentage
1	Zn	60 (min.)
2	Cu	1.7 (max)
3	As	0.1 (max)
4	Cd	0.1 (max)
5	Cr	0.01 (max)
6	Pb	1.25 (max)
7	Ba	0.005 (max)
8	Hg	0.002 (max)
9	Se	0.004 (max)

Work done at NML for MoEF.

CONCLUSION

Most of the pollution control technologies adopted today convert the pollutant from one form to other like in effluent treatment plant aqueous pollutant may be converted into solid waste which is discharged either in open or secured landfill. Both are undesirable from the land point of view, which need to be preserved for future generation. As the leachate generating from the dumping of these hazardous wastes contaminates the environment, efforts must be directed towards not only effectively monitoring and managing the environment but also developing and applying viable and acceptable ecofriendly processes and waste handling/management technologies to cater to the pollution free environmental control of even the small and medium scale industries.

- Main source of raw material for secondary zinc plants in India is imported ash/dross as they are not available indigenously. Thus all those units having proper pollution control devices, eco-friendly process technology and legal permission from the Ministry of Environment and Forest (MOEF) can import zinc bearing raw material. From environmental point of view secondary zinc recovery process is technologically acceptable in India. Basel ban restricted the import of zinc ash, skimmings, drosses contaminated in plenty with several toxic elements.
- Various processes developed indigenously for secondary raw materials are running successfully at different parts of the country. There are some processes ready for commercialization and some are in developing stage.

- Collection and recycling of spent acid battery is the major cause of concern, which has not been achieved to a satisfactory level as compared to developed nations. A legislation as well as appropriate monitoring are to be evolved and enforced for a systematic collection and it's recycling in the units having eco-friendly and waste management process technologies. The slags produced from the primary and secondary lead processing industries must be disposed off to the sites as per specification/instructions laid down by the Pollution Control Boards. Development of low cost and eco-friendly technology should be encouraged in order to sustain the recycling operation, which will not only meet the increasing demand of lead but also conserve the raw materials.
- For battery recyclers, pollution control boards must organise awareness programmes with respect to handling, transport and recycling of battery scrap.
- The existing wastes generated and accumulated from various nonferrous process industries over a long period should be characterized (TCLP and water extract) for its hazardous or non-hazardous nature and a procedure has to be evolved in consultation with industry and CPCB/SPCB/different laboratories for its safe disposal.
- In addition to increased utilisation of solid wastes, efforts are needed towards waste minimisation by attacking the problem at its source, which not only maximise production and product quality, but also minimise the overall environmental impact by: technology upgradation/use of eco-friendly technology, improvement in the quality of raw material, audit on waste minimisation, adopting innovative management strategies, conducting employee awareness programme.

ACKNOWLEDGEMENTS

The authors wish to thank the Director, NML, Jamshedpur for giving permission to publish the paper and the Ministry of Environment and Forest (MoEF), New Delhi for financial support. Thanks are also due to the management of various industries for supplying the required information.

REFERENCES

[1] Raghavan, M. (2001) Secondary recovery of lead and zinc, *Minerals & Metals Review*, 27(2), 28-34.

[2] Sahu K. K., Agrawal A., Pandey B. D. and Premchand (2001), Secondary zinc industries in India problems and prospects., *Mineral and Metals Review*, 27(2), 125-132.

[3] Manhan, S. E., (1994) *Environmental Chemistry*, 6th Edition, Lewis Publishers, Tokyo, 546.

[4] Thadani, B. C., 1996. Challenges and opportunities for Indian Lead, *Mineral and Metals Review*, 58-68.

[5] Sahu K. K., Agrawal A., Pandey B. D. and Premchand (2001), Secondary zinc industries in India problems and prospects., *Mineral and Metals Review*, 27(2), 125-132.

[6] *http://www.cleantechindia.com/eicimage/2102_68/LRE.htm*, July 25, 2003.

[7] Zunkel, A. D. (1997)Review of electric arc furnace dust treatment processes and environmental regulations, *R' 97, Int. Congress Proc.*, Feb. 4-7, Geneva Switzerland, V129-V137.

[8] Reuters news services, *November 4, 2003, http://www.recyclingtoday.com/news/news.asp?ID=4890&SubCatID=74&CatID=23.*

[9] Phugzhenthy, L., Indian non ferrous metals industry-an era of consolidation begins, National Seminar on non ferrous metals- *Stageties for competitiveness and Globalization, NF-2005 towards global Excellanc*e, 22-23 sept, 2005, New Delhi, pp 22-23.

[10] Bhatt Dhimant, *http://blonnet.com/2004/04/07/stories/2004040700331200.htm*, April 2004.

[11] Thomas, George, Binani Zinc Ltd., Kerala, 1998. *Personal Communication.*

[12] Sharma, S. K., *Proc. on Base Metal Tech.*, 8-10 Feb, 1989, Jasmshedpur, 161-176.

[13] Mackenzie, John and Smith, M. R., 18-20 Nov. 1981. The application of imperial smelting process in India, *Pb-Zn-Cd International Symposium*, C11-C34.

[14] Sharma, R. S., HZL, Udaipur, 1998. *Personal Communication.*

[15] Solanki, S. and Singh, Vakil, 1981. The electrolytic zinc plant of HZL at Debari and its development, *Pb-Zn-Cd International Symposium*, E92 - E112.

[16] Bhatnagar, Deepak and Jancy, A., Jan. 29 - 30, 1998. Cobalt recovery from metallurgical waste of zinc industry, *Environmental Waste Management*, NML, Jamshedpur, 230-241.

[17] Basu, K. and Swarnkar, S. R., 7 - 9 Feb., 1990, *Proc. of Hydromet.-90*, Udaipur, 6.14.1 - 6.14.7.

[18] Jaju, N., Sunrise Zinc, GOA, 1998. *Personal Communication.*

[19] Agrawal, K. C., 17-18 Dec. 1998. *National Seminar on Lead and Zinc recycling-Technology and Environmenta*l, New Delhi.

[20] Phugazhenthy, L., Generation and Recycling of zinc wastes from the galvanising industry- Current status and future trends, *National conference on lead and zinc recycling technology* 2&3 Dec., 2004, New Delhi.

[21] Siroya G L, and Ramachandran T C (2000), Zinc recycling in India- An overview, *MMR*, 11, 40-44.

[22] Barcza N. A., Rpbertson. D. G. C., Schoukens A.F.S., Shaw F. Denton G M., Worcester T. W. and Bailey D. J.(1995), Enviroplas technology for the recovery of lead and zinc from lead blast furnace slag". Presented at Recycling lead and zinc into the 21st Century: *6th International Conference of the International Lead and Zinc Study Group*, Madrid, Spain, 18-23 June, 73-89.

[23] Schoukens A S F., Denton, G.M. and Jones, R.T., Pilot-plant production of Prime Western grade zinc from lead A.F.S blast-furnace slags using the Enviroplas process. *Proceedings of the third international symposium on recycling of metals and engineered materials*, Point Clear, Alabama, 12 to 15 November 1995. Publication of the Minerals, Metals and Materials Society, 1995, pp. 857-868.

[24] Premchand, D Jha and K M Godiwala, Recovery of zinc from zinc dross, *National conference on lead and zinc recycling technology and environment*, 2&3 Dec., 2004, New Delhi.

[25] Nogueria E. D., Regife J. M. and Arcocha A. M., winning zinc through solvent extraction and electrowinning, *Engineering Mining,* 180(10), 1979, 92-94.

[26] Nogueria E. D., Regife J. M. and Blythe P. M., Zincex- The Development of a Secondary Zinc Process, *Chemistry and Industry*(2)(1980. 63-67.

[27] Nogueria E. D., Regife J. M. and Viegas M. P., Design Features and operating experience of the Quimigal Zincex plant, *Chloride Electrometallurgy*, ed. P D Parker, Warrendale, PA: The metallurgical society 1982, 59-62.

[28] Martin San Lorenzo; Daniel Regife Vega; Jose M. Nogueira; Eduardo D.(1983), Process for the production of electrolytic zinc or high purity zinc salts from secondary zinc raw-materials, *U S Patent 4,401,531*, August 30.

[29] Lightfoot, B.W., Short, W.E. and Guorgi, G.A., EAF dust processing with Ausmelt Technology, presented to: *The extraction and application of zinc and lead.* Sendai, Japan, May 1995.

[30] Olper, M(1993) The EZINEX process – A new and advanced way for electrowinning zinc from a chloride solution, *International – World Zinc 93*, Hobart, Oct. 10-13.

[31] Olper, M (1994), The EZINEX process – *Proceeding EDP Congress TMS annual meeting*, San Francisco, 513-524.

[32] Cole, P. M. and Sole, K. C (2003), Zinc solvent extraction in the process industries, *Mineral Processing and Extractive Metallurgy Review*, 24(2), 91 – 137.

[33] Hartwell T, "Sustainability – Steel and the Environment" Int conference organized by the Institute of Materials, Minerals and Mining on 1 – 2 November, 2004 in Birmingham, UK, *http://www.iza.com/zwo_org/ZincNews/ZNmenu-F-g.htm#15)*

[34] Pillai Venugopal . *http://www.projectsmonitor.com/detailnews.asp?newsid=*7760.

[35] New lead mine opens in WA, *http://www.abc.net.au/rural/content/2005/s1472012.htm, Friday, 30/09/2005.*

[36] Premchand and Rakesh Kumar, *National Conference on Lead & Zinc Recycling – Technology & Environment*, Dec. 98, New Delhi.

[37] Behrendt, H., Umwelttechnic, M. R., & Freiberg, R'97 *Congress Proceedings Vol. I, p. 34*, Pub. By EMPA, Lerchenfeldstrasse 5, 9014 Saint Gall, Switzerland, 1997.

[38] Viswanathan, S. and Thamhane, A. R., *National Seminar on Lead and Environment*, Indian Lead Zinc Information Centre, Feb. 91, p. 69.

[39] Queneau, P. E. and Siehmund, A., (1996) Industrial scale Lead making with the QSL continuous oxygen converter, *JOM*, 48(4), 38-44.

[40] Mackey T.S. and Perngaman, R. D., (1990), *Lead-Zinc'90*, TMS-AIME, 13-38 & 453-548.

[41] Sahu, K. K., Agrawal, A, Pandey, B. D. & Premchand (2000) Lead industries and Environmental Pollution in India, *Proceedings of the International Conference on Environmental Management in Metallurgical Industries*, Ed. R. C. Gupta, 14-16 December, Institute of technology BHU, Varanasi, 125-132.

[42] Prengaman, R. D. and McDonald, H. D., (1980), *US Patent, 4229271.*

[43] Cole E. R., Lee A. Y., and. Paulson D. L., Electrowinning of lead from H_2SiF_6 solution, *US pat No. 4,272,340*, June 9, 1981.

[44] Olper M., *CX-EW Process - A comprehensive recovery system for lead acid battery, International Pb-Zn Study group*, Rome (11-13 June 1991), 79-90.

[45] Reynolds R. M., Hudson E. K., Olper M, *Lead-Zinc 90*, ed. T. S. Mackeyand R D Perngaman (Warreldale, PA, TMS, 1990), 1001-1022.

[46] Ginatta M V, *US Patent 4451340*, May 29, 1984.

[47] Moure, W. B. and Etsell T. H., *US Patent 5211818* (May 18, 1993).

[48] Bratt G. C. and Pickring R. W., (1970), Production of lead via ammoniacal ammonium sulphate leaching, *Met Trans*, 1, 2141-2149.

[49] Spijkerman J. B. and Groenen, R. J., (1991), *U.K Patent application*, 91029942.

[50] San Lorenzo D. M. and Nogeuira,G. D., (1991), *Hydrometallurgical Treatment of lead secondaries: The Placid process*, Interntionl Lead zinc study group, Rome (11-13 June), 315-336.

[51] Frías, C., García, M. A., Díaz, G., 2000. New clean technologies to improve lead-acid batteries recycling. In: *Proceedings of the Lead–Zinc'2000 symposium*. TMS, Pittsburgh, 791–80

[52] Díaz, G., Martín, D., Frías, C. and Sánchez, F., (2001) Emerging applications of ZINCEX and PLACID technologies. *Journal of Metals,* 53 2, 30–31.

[53] Vincent Rich, *The International Lead Trade*, Woodhead Publishing Ltd., UK, p. 103, 1994.

[54] Rajagopal R, Tharian. J, and Harmati. Z, EMPA St. Gallen, Switzerland, Recycling in India - Environmental and Technological Imperatives, Paper presented at the R´99 Congress (Recovery, Recycling, Re-integration), February 1.999(*http://www. environmental-expert.com/articles/article428/article428.htm*).

[55] Edward Worden, AMM, Dec 27, 2000, *http://www.findarticles.com/cf_dls/m3MKT/ 248_108/68769276/p1/article.jhtml*

[56] Narayanan, P.A.I and Mathur G. P.(1984), *Zinc oxide from zinc ash/zinc hydroxide*, Internal report, National Metallurgical Laboratory.

[57] Chandra N, Bhasin S, Sarkar D and Sharma M, Solid waste minimization in secondary zinc industry, *National conference on lead and zinc recycling –technology and environment*, 2&3 Dec., 2004, New Delhi.

[58] Basak G, Lahari R N and Sahani M J(1981), Process parameter of zinc dust production and application technology, *Pb-Zn-Cd International Symposium*, 18-20 Nov, New Delhi, 307-340.

[59] Lahari R N (1980), Distillation of zinc: PartI Vacuum distillation of zinc, *Trans. IIM,* 33(4), 322-326.

[60] Basak G(1980), Distillation of zinc: PartII Study on lead constituent in distilled zinc, *Trans. IIM*, 33(4), 327-328.

[61] Thangaraj T, Tenth five year plan report, Lead recycling-Trends & Development, *National conference on lead and zinc recycling –technology and environment*, 2&3 Dec., 2004, New Delhi.

[62] Jagannathan, V., 18-20 Nov. 1981. Lead smelting operations at HZL Vizag, *Proc. on Pb-Zn-Cd, Int. Symp.*, E50 - E72.

[63] Mukherjee, P., 18-20 Nov. 1981. Lead smelting operations at HZL Tundoo, *Proceedings on Pb-Zn-Cd, Int. Symp.*, E1-12.

In: Solid Waste: New Research
Editors: A. S. Martin and J. D. Suarez

SBN: 978-1-60456-809-7

Chapter 3

An Economic Analysis of Municipal Solid Waste Management and Policies

Francisco J. André*[1] and Emilio Cerdá[2]
[1]Universidad Pablo de Olavide
[2]Universidad Complutense de Madrid, Spain

Abstract

This work presents an analysis of municipal solid waste management, from an economic perspective. We discuss the market failures caused by generation and management of waste and the main policy instruments to deal with those failures. Specifically, we focus on incentive policies and we list the main advantages and disadvantages of each instrument. We also present a brief summary about the recent history concerning both the principles and practices for waste management in the international context.

Keywords: municipal solid waste, market failures, incentive policies, landfilling, recycling.

Introduction: General Issues on Waste

Waste is usually defined as stuff we do not want –and hence we are willing to pay to get rid of it. In economic terms, we could say that waste has a negative value or, in other words, it is a *bad* rather that a *good*. Classifying residuals is not an easy task because very different criteria may be used, according to waste hazard, origin, composition or treatment options. A first important distinction should be made between hazardous and non-hazardous waste. A more detailed classification can be made based on waste origin: mining waste, agricultural waste, hospital and medical waste, radioactive waste, industrial waste and municipal solid waste (MSW).

---

* Corresponding author. Departamento de Economía, Universidad Pablo de Olavide, Carretera de Utrera, km. 1, 41008, Sevilla. Spain. Phone: 954349120. E-mail: fjandgar@upo.es

In this study, we focus on the latter, MSW, which typically includes the trash of households and small business, although the specific elements included or excluded in the definition depend on the law of each country or region.

The generation, management and public policies related to MSW are attracting increasing attention from many points of view. This is crucially motivated by the increasing flows of MSW, as a consequence of the process of economic growth and development, the changes in consumption patterns and the population shift from rural to urban areas. The latter demographic trend has also created the need to look for sustainable management strategies, including sensible waste management practices. Figure 1 illustrates how municipal waste generation has grown (and is expected to grow) roughly at the same speed as Gross Domestic Product (GDP) and faster than population, implying an increasing per capita generation rate.

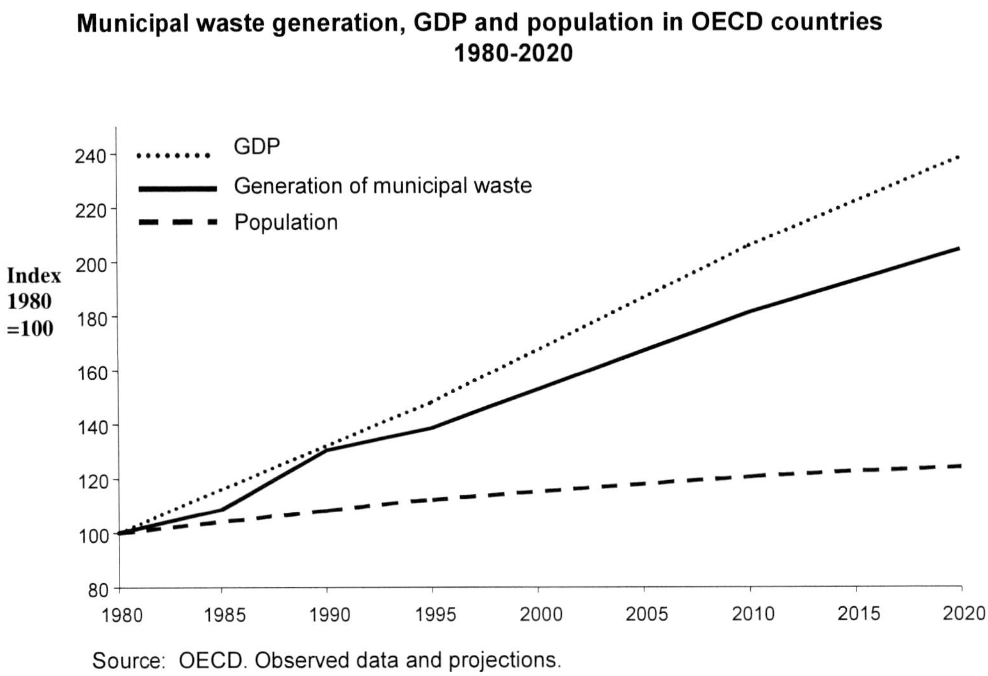

Figure 1.

Another reason to focus on MSW is that, unlike other types of waste such as industrial or radioactive, which can only be handled by highly specialized staff, many people are directly or indirectly involved in MSW, since its generation, treatment and management strongly depends on the participation of citizens. The provision of incentives for consumers and firms to act in socially desirable ways when dealing with MSW also opens up a fruitful field for the application of public policies.

This work aims at offering a revision of all the economic aspects of MSW, focusing on the management strategies, the market failures arising from MSW and the public policy instruments available to deal with those failures. We try to combine all these aspects in a coherent way and, at the same time, provide a broad survey of the recent related economic literature, which can be basically located in the field of Environmental Economics, since most economic issues on MSW have to do with its environmental impact.

Although this study mainly focuses on economic aspects, it is convenient to bear in mind that MSW management involves a complex combination of legal, technical and environmental (not to mention economic) elements. Quadrio-Curzio et.al. (1994) offer a collection of multidisciplinary studies analyzing different aspects of MSW management in Europe. In the introduction of that book, it is stated that "the technological literature on MSW is indeed vast, but is often centered on very specific problems and is sometimes highly technical in nature. The environmental literature is immense but disperse, and mainly focuses on specific cases. The economic literature internationally available on MSW is much less developed and it is only in recent times that economists have come to be involved in the analysis of the MSW problem" (p. vii). Beede and Bloom (1995) discussed a number of current issues related to MSW. Halstead and Park (1996) offer an overview of those aspects related to economic policies and the role of government. A recent update of the economics of waste can be found in Porter (2002).

The structure of the remainder is the following: Section 2 begins by presenting a synthesis of the fundamental economic aspects in MSW management. In Section 3 the main market failures related to MSW management are discussed. The main economic policy objectives and instruments are presented and classified, giving most attention to the what are called economic incentive instruments. Section 4 presents a synthesis of the recent evolution in the management and policies of MSW at an international level, highlighting the facts which have produced the present situation and the present trends. Finally, a resumé about the present situation a trends of MSW management in Europe, the United States and Japan is given. Section 5 contains the main conclusions of the work.

MSW Management

MSW management can be defined as the set of activities needed to collect, transport, pack, store, transform and eliminate the waste generated in a given area in an environmentally, economically and legally suitable way, according to their properties, volume, origin, recycling possibilities and costs. MSW management can be divided into basically four different stages: pre-collection, collection, transport and treatment.

Pre-collection refers to storage, manipulation, classification and placement of waste in the right places and times to be collected and transported. The relevance of this stage for the correct working of the following has lead the municipalities, in recent times, to spend a great deal of effort to improve it by installing selective containers and making campaigns for public participation.

The stages of *collection* and *transport* are usually the most costly of the whole process and they require detailed planning, including the frequency, timing and itineraries of collection, staff, equipment and financing. The main environmental implications of this part of the process basically depend on the vehicles used for the transportation and the loading and unloading operations. Waste can be transported directly to its final treatment or disposal facilities or to the so-called transfer plants, where it is packed and loaded in bigger trucks, more suitable for the transportation to its ultimate destination.

The latest part of the management process is the *treatment* stage, which includes the operations needed to eliminate the residuals or to exploit the materials contained in them. The most important (legal) treatment methods are landfilling, incineration, recycling and composting. In the following pages, we will briefly address the main economic and environmental features of each of them:

Landfilling

The first, and traditionally the most widely used, waste treatment method is the disposal of waste in a landfill. The rest of the methods typically cannot eliminate all the waste, but they always generate some by-product that needs to be disposed of, normally, in a landfill.

The main environmental risks associated to landfills are due to the leachates, which are the liquids resulting from the decomposition of waste. These leachates may seriously pollute the soil as well as surface and underground waters. There are also other potential risks linked to uncontrolled explosions because of methane gas coming from the fermentation of organic waste products, as well as unpleasant smells, and pollution by light elements (plastic bags, powder, pieces of paper…) and the proliferation of insects and rodents.

Economically, a landfill can be approached from several points of view: first, from the perspective of citizens living close to it, a landfill can be seen as a negative externality or disutility, which is not compensated for in the market. It may be claimed that this disutility is already internalized in the price of houses close to the landfill, which is expected to be lower.[2] Nevertheless, this argument only applies to those people who decided to buy their houses aware of their locations and the presence of the landfill, but not to those who bought a house before the landfill was located and suffered the location afterwards. On the other hand, the price of houses, in the best of cases, may reflect the effect of expected damages, but will not typically reflect the effect of unexpected accidents and security failures. The fact that citizens are typically unwilling to accept that noxious facilities are located close to their houses is a traditional issue in the literature and it is usually referred to as the NIMBY ("Not in My Back Yard") syndrome.

Secondly, from the point of view of the firm or agency responsible for waste management, constructing a landfill can be seen as an investment, since landfilling is an alternative for waste treatment, so that using a landfill allows cost savings in other treatment techniques. The related decisions involve comparing the cost of the different treatment alternatives. Landfills are characterized by large initial set-up costs, including the cost of land, the construction costs and the preparation of the landfill to satisfy all the safety, health and environmental requirements. The main operating costs during the lifetime of the landfill are associated with collection and transportation of residuals. Finally, when the capacity of the landfill is depleted, it must be closed and some further costs must be incurred to guarantee the safety requirements

Finally, landfills can be seen as natural resources, where the capacity of the landfill plays the role of the "stock" of the resource and the landfilling rate can be seen as the extraction rate of the resource. When seen as natural resources, landfills present some interesting

---

2 Nelson, Genereaux and Genereaux (1992) estimate a negative effect of the proximity of a landfill on the prices of houses in Minnesota.

specific features: Hanley et.al. 1997 (chapter 8) defends the classification of natural resources as renewable and nonrenewable, as being more accurate than the traditional classification of exhaustible and non-exhaustible. Although this is a sensible claim in general, it may not fully apply to landfills. Land has some assimilative capacity that can be considered as renewable, (see, for example, Pearce and Turner, 1995). Nevertheless, the amount of generated residuals per unit of time in urban areas greatly exceeds the assimilative capacity of land which, in practice, becomes irrelevant, since the capacity of any landfill will be depleted sooner or later. Henceforth, in the case of landfills, the most suitable category seems to be that of *exhaustible resources*. Nevertheless, as noted in Ready and Ready (1995), unlike other resources, whose depletion is irreversible, once a landfill is full, it can be replaced by a new one, so that they can be categorized as *exhaustible and replaceable resources.*

On the other hand, typically the "stock" and location of resources is exogenously given by nature, but in the case of landfills, both the stock (i.e., the capacity) and the location are economically relevant decision variables. André and Cerdá (2001, 2004) study the problem of optimally determining the capacity of a sequence of landfills, whereas André, Velasco and González (2004) analyze the joint capacity and location problem. Jacobs and Everett (1992) present a model to determine the optimal use of several landfills when there is the possibility of recycling. Ready and Ready (1995) study the optimal price policy of landfills in a dynamic context. Huhtala (1997) determines the optimal switching time from an old landfill to a new landfill. Gaudet, Moreaux and Salant (1998) present a spatial and dynamic model to analyze the transportation of residuals from several cities to several landfills. Ley, Macauley and Salant (2000, 2002) apply the latter work to evaluate the economic costs of restricting waste trade among states.

Incineration

Incineration is a controlled combustion process that transforms the organic material contained in waste into inert materials (ashes) and gases. Incineration is quite widespread in Europe and other parts of the word, such as Japan, where land is very scarce, so that landfilling can be comparatively very expensive. Incineration allows the weight (around 70%) and volume (from 80% to 90%) of waste[3] to be reduced very quickly, without the need to store residuals for long periods of time, in landfills or deposits. Furthermore, it avoids the consequences of organic decomposition since ashes are inert materials. The facilities needed for incineration require little space as compared to landfilling and they can be placed in any type of location, even in urban areas. By the combustion of some materials, notably plastics, which present a higher heat power, it is possible to get electricity or heat as a by-product, which allows other energy sources to be saved.

As for the disadvantages, this method never enables all the residuals to be eliminated, because not all the materials can be burnt, and even those which can be incinerated generate ashes as a residue that cannot be reduced any more and require some complementary method (typically a landfill) to get rid of it. Some complementary method is also needed to deal with a situation of breakdown in the incineration plant.

---

[3] See Brunner (1994).

From an economic point of view, it is a costly method, because incineration facilities require a high initial investment for the installation, and high operating costs, since the operation technique is very highly specialized and always requires some external energy source. Some highly specialized method is needed to deal with the potentially very harmful atmospheric emissions from incineration, mainly including dioxins and furans. This potential risk of incineration has made it particularly unpopular among ecologist groups. Finally, incineration is irreversible as it totally destroys the incinerated materials.

The economic decision to set up an incineration plant involves forecasting the amount of waste to be incinerated, the kind of residuals, their heat power and the price of the energy to be obtained. Since it is a very capital-intensive alternative, it exhibits increasing returns to scale and it is only profitable above a given amount of operations. Keeler and Renkow (1994) present a theoretical model to analyze incineration as compared to landfilling and recycling and they study decision of setting an incineration plant or not.

As with landfilling, the setting up of an incineration plant is subject to the NIMBY syndrome, as it generates adverse reactions in the neighborhood and typically causes the prices of surrounding houses to drop. Kiel and McClain (1995) study the effect of setting up an incineration plant on the prices of houses before the construction, while the rumor of the installation is being spread, during construction itself and when the plant is already working. They find a negative effect on prices during a period of seven years. Kunreuther et.al. (1987) propose a compensation mechanism for the installation of noxious facilities using an auction system. Ye and Yezer (1997) analyzed the location decision of this kind of facilities (landfills or incinerators) considering two economic effects: the scale economics and the vicinity to the affected neighborhoods.

Recycling

Recycling allows to part of the goods or materials from the waste flow to be transformend and reincorporated into the production process. Recycling has recently become increasingly popular because of its environmental and economic advantages, which can be grouped in two basic categories: first, using recycled products instead of virgin materials allows the saving of scarce resources (see Weinstein and Zeckhauser (1974), Schulze (1974), Lusky (1975a,b), Hoel (1978), Dinan (1993), Huhtala (1994)). Secondly, recycling as a waste treatment technique is more environmentally friendly than alternatives such as landfilling or incineration, since it generates less pollution and saves landfill space. (see Plourde (1972), Smith (1972), Lund (1990), Sigman (1995), Ready and Ready (1995), Highfill and McAsey (1997), Huhtala (1997, 1999)).

In the last decades, the interest for the first advantage has been moderate because of the discovery of new mineral stocks and the technological development which increasingly allows us to reduce resource scarcity and to replace part of the resources for man-made capital. On the other hand, increasing pollution and the scarcity of land available for landfill construction has greatly dented the interest for the second aspect. An additional advantage of recycling is its reversibility since it does not imply fully destroying the involved products and materials (as is the case with incineration). Huhtala (1999) studies the optimal allocation of resources between two technological sectors: regular production and recycling. André and Cerdá (2005) stress the technological nature of recycling and analyze the effect of this technique on the overall production possibility of the economy, as well as the optimal decisions about production and resource use. Duggal et. al. (1991) study the economic mechanisms motivating the households' decisions about the amount of recycling effort. Di Vita (2001) studies the connection between recycling, resource use and sustainability in an endogenous growth framework.

Some authors have pointed out that the popularity of recycling may induce an over-utilization of this method beyond the economically reasonable limits, because of public opinion and image questions. Along with its advantages, recycling also presents some disadvantages and limitations. As noted in Williams (1998, chapter 4), the economic viability of recycling depends on the following aspects: first, a stable and risk-free provision of suitable waste materials; second, an adequate system to collect and transport materials to the recycling plant. The need for a selective collection implies that collection and transportation costs are somewhat higher than those of other methods; third, a reliable system to clean and classify materials to be processed and used again in the production process; and fourth, stable and secure markets for recycled materials.

Hoel (1978) explicitly models the environmental costs of recycling. Baumol (1977) studies the viability of recycling to solve waste management problems. Chilton (1993) underlines the need to design the MSW management policy according to rational criteria rather than emotional motives.

Pearce and Brisson (1994) propose a framework to establish the optimal recycling level versus waste disposal according to the traditional marginalist condition equalizing marginal cost and marginal benefit. Recycling benefits include, on the one hand, the generation of recycled products (which can be valued using its market price) and, on the other hand, the saving of (environmental and economic) disposal costs. These benefits should be compared to recycling costs, which include those of selective collection, those of material processing and those from the environmental impact of recycling (if any). If we make the simplifying assumption that recycling has no environmental impact, the optimal recycling level can be identified as that value that satisfies the following condition:

$$P_R + C_E + C_{EA} = C_{RS} + C_R$$

where P_R stands for the price of the recycled material, C_E (C_{EA}) represents the economic (environmental) marginal cost of waste disposal, C_{RS} is the marginal cost of selective collection and C_R stands for the operating marginal costs of recycling. Rearranging this equation, we get $P_R - C_{RS} - C_R = -(C_E + C_{EA})$, or alternatively,

$$\Pi_R = -(C_E + C_{EA})$$

where $\Pi_R \equiv P_R - C_{RS} - C_R$ represents the private marginal benefit of recycling. Pearce and Brisson (1994) propose two alternative policies to induce the socially optimal solution: the first is a recycling subsidy equal to C_E+C_{EA}, and the second is a recycling subsidy equal to C_E and a tax on waste disposal equal to C_{EA}. The available policy methods are discussed in more depth in section 3.

The private (market) solution and the social optimum are illustrated in Figure 2. A profit-maximizing firm will choose the level of recycling R_1, which ensures $\Pi_R = 0$. This is the private solution or *free market optimum*. The level R_2 can be labeled as *financial optimum*, since it is the result when the saving in disposal costs C_E is taken into account but the environmental cost is disregarded. This recycling level can be reached if the recycling firms receive a subsidy of C_E monetary units for each recycled unit. If, moreover, the firm is subsidized for the saving in environmental costs, C_{EA}, the social optimum R_3 can be reached. Any recycling level above R_3 would be inefficiently high.

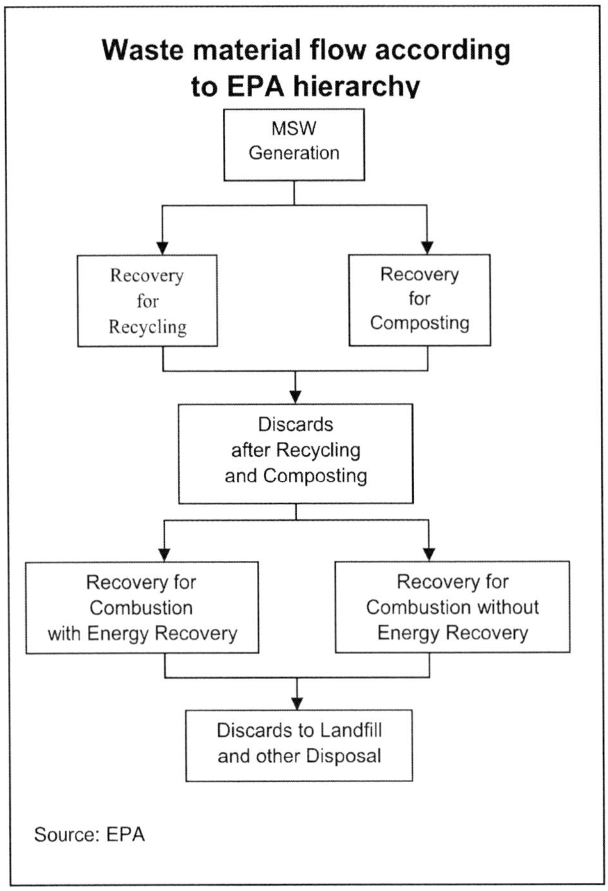

Figure 2.

Composting

Composting is an anaerobic biological decomposition process of organic material contained in waste under controlled conditions. The process involves the manual or mechanic separation of metal, glass and plastic contained in the residuals, to isolate the organic elements. The duration of the composting process is determined by the fermentation of the materials. This fermentation can be done in the open air (taking about three months) or in special digestors (taking about fifteen days).

This technique allows the recuperation of the organic material to be used in agriculture. Composting generates as output an organic fertilizer or *compost*, which is very rich in nutrients and produces very beneficial effects on the soil, such as regulating the compactness of land, helps chemical fertilization, increases the ability of the soil to keep water, provides nutritive elements and increases the organic contents of soil. Moreover, by reducing the organic contents of waste, it also alleviates the biodegradation process and the generation of leachates in landfills.

Since composting allows part of waste to be recovered and can be seen as a particular type of recycling. Nevertheless, it also presents some specific features distinct from recycling: first, composting is suitable only for organic material, but it allows the combining of different organic materials to generate compost. On the other hand, recycling requires a particular technique for each specific material such as glass, paper, plastic or metal. Secondly, while recycled products sometimes can be used again in the same activity as the original one, the output from composting is always used for a different use than the original one and it is limited to agriculture.

Reuse and Source Reduction

Besides what we can call regular waste management methods (those described up to now) it is common nowadays to refer to other practices such as reuse and source reduction. Although these are not correctly speaking treatment or management practices, they are currently considered as crucial strategies in the field of waste management. As a matter of fact, these are considered as key strategies and governments are strongly promoting them, following the maxim that "the cheapest way to manage waste is not generating waste".

Reuse means using some good or element more than once, in the same or in different activities, without being processed or transformed. This option is typically used for items such as plastic bags, glass bottles, cupboard boxes or some car components. Reusing a product extends its lifetime and reduces the waste flow. Moreover, if replacing a new product by a reused one allows the saving of the production cost of the new one, which includes the requirements of energy and materials. Nevertheless, a subtle trade-off should be underlined here: a product that is to be used more than once needs to be more robust, i.e., a thicker and heavier format, and henceforth, it requires a larger amount of resources.

Source reduction is a wider concept which includes any practice or strategy to modify the production and consumption patterns in the sense of minimizing the amount of waste to be treated. In this sense, as reuse helps in reducing the amount of waste, it can be seen as a particular case of reduction.

Concerning the limitations of source reduction, since it is a rather recent trend, there is great uncertainty and lack of information about its viability and practical implementability. The changes to reduce the waste flow largely depend on the existence of alternative (less waste-intensive) products or production techniques. For example, in some cases, it is possible to reduce the packaging of goods, but this option may not be viable for all the goods because of conservation and safety reasons.

Moreover, even if it is technically possible, the success of source reduction strategies typically depends on the willingness of firms and consumers to change their production and consumption patterns. As far as firms are concerned they will probably be reluctant to change their production methods if this implies a substantial cost increment. Consumers, in turn, may perceive a welfare reduction if they have to change their consumption decision from those preferred. In this context, public policy plays a crucial role to provide the right incentives for firms and agents to change their habits. This role of economic policies are studied in Jenkins (1993), Goddard (1995) o Palmer, Sigman and Walls (1997) among others.

Integrated Waste Management Hierarchies

In general, it is not possible to identify a single waste treatment option as unambiguously superior to all the rest or discard any option in all the cases. Source reduction is a valuable strategy to reduce waste management costs, but it has a limited scope and, beyond some threshold, an additional reduction may imply large welfare reductions. Recycling and composting allow the recovering of some materials, but some elements cannot be recycled or composted, while others are inefficiently costly to be recycled or composted. Incineration can reduce the volume of waste, but it generates a by-product that can not be eliminated and has some potentially very harmful environmental implications. The use of landfilling may be unlimited in principle, but an excessive use of landfilling can be environmentally and economically inefficient. Typically, an optimal management strategy implies combining different methods.

The so-called *integrated waste management* refers to a rational combination of different methods, ordered according to some economic and environmental criteria. The use of these criteria to order treatment methods gives rise to a so-called hierarchy. A well-known hierarchy is the one proposed by the US Environmental Protection Agency (EPA), which follows this pattern:

1. Source reduction
2. Recycling
3. Incineration (if possible, with energy recovery) and landfilling.

According to this hierarchy, waste goes through the stages shown in Figure 3: The first option is to reuse and reduce waste when possible. When this option is not feasible any more, the next preferred option is recycling and composting. Finally, that waste that can not be recycled or composted must be treated with the least preferred methods: incineration (if possible, with energy recovery) and landfilling.

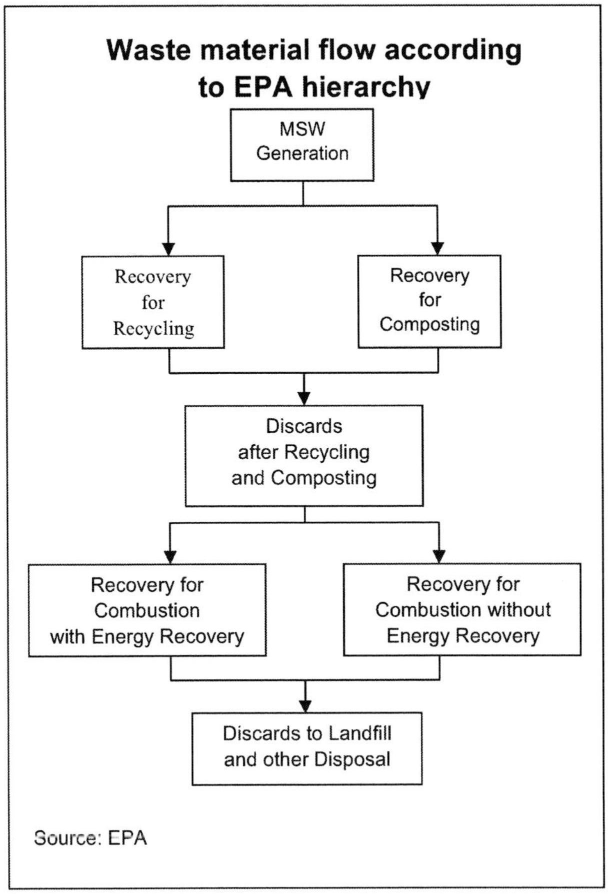

Figure 3.

A Broader Version of Waste Management

In the previous section, we have followed a conventional approach to waste management, which we could label as "post-consumption". This approach basically involves taking the waste volume and composition as given and finding the most suitable composition of treatment methods for it. Applications of this approach can be found, for example, in Lund (1990), Jacobs and Everett (1992), Keeler and Renkow (1994) or Huhtala (1997).

Nevertheless, taking into consideration some important aspects, such as source reduction, requires a more comprehensive vision, which we can call "pre-consumption". According to this approach, the actions needed to manage waste in a suitable way start in the previous stages of production and commercialization of consumption goods, because many decisions made at these stages determine waste volume and composition, and hence, they are essential for the treatment options. This approach is adopted, for example, in Dinan (1993), Morris and Holthausen (1994), Atri and Schellberg (1995), Fullerton and Kinnaman (1995) or Huhtala (1999).

This approach is also more suitable to describe the connections of production, waste management and the environment: production requires natural resources, which are transformed into consumption goods, then into waste and, if some recycling or composting technique is used, a part of these resources can be recovered, whereas the other part is turned into pollution. These relations are illustrated in Figure 4.

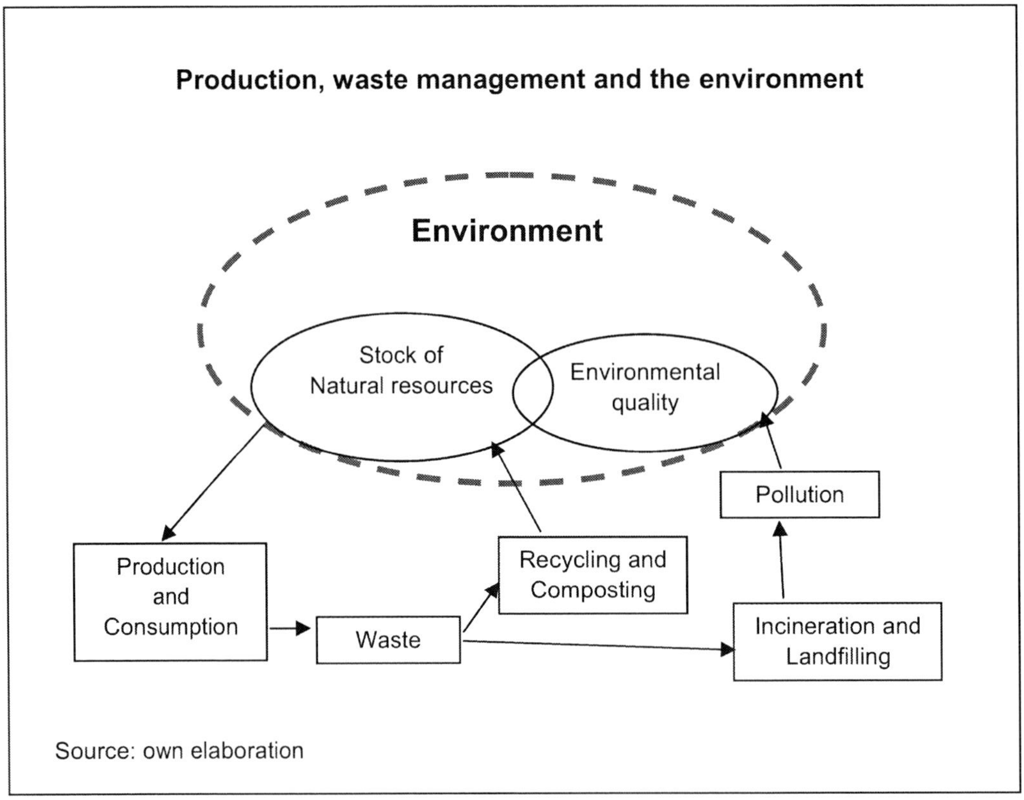

Figure 4.

Pearce and Brisson (1994, p. 213) claim that the following sequence of "right questions" should be answered to develop a waste management strategy: «The first question is how much waste there should be. This involves a balance between source reduction, recycling and final disposal. Little of the current policy discussion focuses on this issue: instead, at best, it asks what should be done with the waste we have, not how much there should be. Once this question is answered, the next stage is to decide between disposal and recycling for the waste that does arise. Again, there is confusion mainly arising from a presupposition that recycling is intrinsically good and/or socially profitable. But recycling uses up resources and generates its own waste problems, so the issue must be the optimal balance between recycling and disposal. (…) Finally, the last question is how to achieve the desirable balance".

In order to answer the questions formulated by Pearce and Brisson, we should think of a framework like the one displayed in Figure 4, i.e., we should keep in mind the connections between waste management and production and consumption activities. The first question on the optimal amount of residuals should be answered considering the costs and benefits associated with that amount of waste. But these costs and benefits, in turn, depend on the methods used to manage and treat the waste, which is determined in the second question.

To solve this difficulty, we propose an approach based on the Bellman optimality principle,[4] which is the basis for Dynamic Programming. Applying this technique, our proposal is to answer the second question first and, afterwards, answer the first one keeping in mind the effect on the second. Another observation on the framework proposed by Pearce and Brisson is that the waste treatment possibilities do not only depend on the amount of waste, but also on its composition, so that both the amount and composition of waste should be decided in the production stage.

Summing up, we propose the following decision framework: first, assume a generic flow of waste Q, where Q is a vector specifying the amount of each material contained in the waste flow (henceforth, including information on volume and composition). Given Q, find the optimal combination of treatment methods, which gives the minimum cost function $C(Q)$. This cost function should be integrated in the production decisions so that, when deciding what to produce and how to produce it, apart from the regular production cost, the effects on waste management cost should also be considered.

MARKET FAILURE AND PUBLIC POLICIES

Waste and Market Failure

As noted in Zoboli (1994), the generation and management of MSW represents a kind of *cascading external effect* due to the heterogeneity of the economic agents involved and the lack of regulated interaction channels among them. Since the virgin materials are extracted and transformed to produce consumption goods, until the residuals are generated and treated, materials go via different agents in a transformation chain, which is not necessarily articulated in a rational way. Typically, these agents do not have any incentive to consider the effect of their activity on the rest of the agents in the chain. In this framework, suboptimal decisions could be made at any point of the chain.

In the production and commercialization stage, the *volume and composition of waste* is determined. These decisions are typically made without taking into account their consequences on the next stages. Concerning the volume, if there is no mechanism which makes waste reduction profitable for the firm, we should not expect the entrepreneurs to put any effort into source reduction. On the other hand, the material composition of waste may be crucial to determine its recyclability. So-called "green design" or "design for the environment" is a practice which consists of manufacturing products taking into account their environmental impact and future recyclability. Fullerton and Wu (1998) pioneered this line of research in economic models. Eichner and Pethig (2001, 2003) included in an economic model the material flow through the different stages of production, consumption and waste treatment. André and Cerdá (2005) model a production process depending on a recyclable and a non-recyclable resource and study the interaction between regular production technology and recycling technology.

---

[4] Bellman (1957).

Consumers are essential for the *pre-collection stage* of waste. Since separating different materials is costly in terms of time and welfare, they are not expected to devote enough (i.e., an efficient amount of) time to this task if they do not get any incentive for it.

Firms responsible for waste *transport and treatment* may not internalize the environmental effects of their decisions. Recycling may provide some disposal cost saving which accrues to the government or the agency responsible for disposal, rather than to the person or firm who generates the waste (and perhaps decides on recycling or not). Moreover, the environmental benefits of recycling are not typically perceived (at least, not fully perceived) by individual consumers and firms. In this context, recycling effort will typically be inefficiently low.

Policy Objectives and Instruments

As discussed immediately above, waste generation and management is typically subject to several market failures. Correcting these failures provides the main *policy objectives*. These objectives can be grouped into "upstream" and "downstream" depending on which part of the process of waste they refer to. Upstream objectives usually aim at reducing waste generation at the production and consumption stages, as well as promoting the design of more easily recyclable products. Downstream objectives have to do with the right use of waste management methods.

Similarly, *policy instruments* can also be classified as upstream and downstream. A usual discussion in this field is whether downstream instruments (operating directly on the waste management stage), may themselves succeed in attaining upstream objectives (i.e., getting desired effects on the volume and composition of waste). See, for example, Calcott and Walls (2000, 2005) or Choe and Fraser (2001).

Policy instruments can also be classified, according to their nature, in *command and control* and *incentive* instruments. The former involve setting standards and principles to be met by those practices related to waste generation and management, such as the prohibition on using some kinds of materials in production, fixing quantitative recycling requirements or limiting waste trade (see, for example, Ley, Macauley and Salant 2000, 2002).

We focus on *incentive instruments*, which are based on economic objectives rather than on direct command. The incentive or economic instruments can be classified in the following categories:[5]

Taxes Proportional to the Amount of Waste Produced and Disposed of

This is a typical downstream instrument since it is based on waste itself, rather than on production or consumption. Taking waste as a negative externality and considering the amount of waste as a good quantitative measure of this externality is consistent with the setting of a price per unit of waste so that the external cost is internalized.

Applying these "pay-as-you-throw" (PAYT) systems requires the consumers to have to buy an official bag, can or tag for waste to be collected. The tax takes the form of a price paid for each bag ("pay-per-bag"), can ("pay-per-can") or tag ("pay-per-tag"). For some pioneering experiences see Hong et.al. (1993), where a disposal service demand function is

---

[5] See Zoboli (1994).

estimated to value the effects of price incentives. Fullerton and Kinnaman (1996) study the socioeconomic determinants for the consumers to react to one of these systems. Callan and Thomas (1999) analyze the determinants for a municipality to adopt a PAYT system. Jenkins (1993) displays a detailed study of proportional waste taxes.

The main advantage of proportional taxes is that, by internalizing (at least part of) the social cost of waste, they generate an incentive for source reduction.

Some practical shortcomings have also been pointed out: first, taxing the legal alternatives may provide an incentive for illegal disposal. Fullerton and Kinnaman (1995) and Choe and Fraser (1999) study this possibility and the policy instruments to be applied in this case. Another limitation is the difficulty in measuring the amount of waste generated by each household. In densely populated areas, this measurement can be virtually impossible.

Taxes on Goods Packaging

This is an upstream instrument focusing on an initial stage, when materials have not yet been transformed into waste. A different tax is generally set for recyclable and non-recyclable packaging products, so as to place the latter at a disadvantage. The aim is to reduce the proportion of non-recyclable material which is present in the waste flow.

The most common critiques of this type of instruments are the following: first, since packaging is a typical instrument for product differentiation, we could expect the demand for highly differentiated products to be very inelastic, so that, taxes will be almost entirely transferred to the final consumer price without any substantial effect on the amount of packages. Similarly, it is often argued that the income effect of the tax is likely to predominate over the incentive effect. Both of these problems are expected to be particularly severe when the tax is fixed at too low a level compared to the final product price. But, on the other hand, a very high tax rate can also have some other efficiency costs, apart from generating social opposition.

In practice, these kinds of instruments have been widely used in Europe, for example with taxes on beverage containers in Germany and Sweden or on plastic bags in Italy.

Incentives to Recovery and Recycling

This is a downstream instrument focused on reducing the disposal costs, as well as promoting recovery and recycling to substitute for virgin materials in production.

A first example of this group is taxation on virgin materials to promote the usage of recycled material instead. Pearce and Turner (1993) and Bruvoll (1998) defend these taxes based on the argument that free market prices of virgin materials tend to encourage over-extraction, also producing a natural negative externality and too much waste.

Some critiques are also made in the literature on this instrument. Dinan (1993) points out that this instrument is only valid for natural resources with directly recycled substitutes and does not encourage the use of recycled products that do not directly replace a virgin material. Another limitation is that these tributes cannot be designed to reflect the net disposal cost saving, since the marginal cost of recycling is not the same in all sectors that use the recycled product. The alternative proposed is a combination of a disposal tax and a subsidy for the use of recycled materials. A similar conclusion is obtained in Fullerton and Kinnaman (1995).

Deposit-refund Systems

Apart from what we can call “simple instruments”, it is also possible to combine them to obtain more complex instruments. Perhaps the best-known combination is the so-called deposit-refund system, which involves increasing the price of packaged products and returning the difference if the package is returned, so that it amounts to a combination of a tax on packaging plus a subsidy for recycling. This system has been traditionally used in many countries and it has some important advantages, such as very few administrative difficulties and the direct application of the polluter-pays principle. As for the limitations of these instruments, apart from not being applicable to all products, its effectiveness crucially depends on the working of the recycling sector.

For the comparison of different economic instruments, see Dinan (1993), Pearce and Turner (1993), Fullerton and Kinnaman (1995), Sigman (1995), Palmer et al. (1997), Palmer and Walls (1997), Fullerton and Wu (1998), Shinkuma (2003).

Present Situation and Prospects

In Hafkamp (2002) a distinction is made between “old” and “new” regimes, in the management of solid waste, regime being understood as a “collective agreement, expressed or tacit, allowing agents to coordinate their activities” (Boyer and Orléan, 1994).

The *“old” regime* can be identified with the period from midway through the XIX century to the beginning of the seventies. The fundamental value it is based on is public health: urban solid waste had to be removed as a measure to prevent the spread of contagious diseases. It also prevented smells and nuisance caused by solid waste. The principal was accepted that every human being has the right to throw away any rubbish he cares or needs to. The most important characteristics of the “old” regime are the following:

- The municipality is charged with the collection of rubbish as often as is judged necessary, using pertinent means without jeopardising public health, and causing the least nuisance possible. The municipality is obliged to accept all rubbish deposited or handed in.
- Final costs incurred in the services of collection and disposal shall fall on domestic and company economies.
- The most-used treatment method is tipping. Incineration and composting are only used in highly populated areas. In some industrial sectors and some businesses the deposit and refund system is used, but only when it is profitable from the economic viewpoint. Recycling is an activity linked to poverty (some people, to earn a living, work at picking up food waste, papers, old clothing, etc).
- The city council has all the competences in managing urban solid waste.

The so-called *“new regime”* began to emerge in the 70s, and produced significant changes from the previous situation. It is not a closed system, rather, as pointed out in Hafkamp (2002), it is “important orientation guidelines”, or general lines which are being implemented at the present time but which are still in the process of reaching maturity. Their principles and basics stem mainly from environmentalists and activists in the fields of the environment and energy use. The fundamental value they are based on is responsible management of natural and environmental resources. Public health is not forgotten (this was a fundamental value in the "old regime"), but, rather, it is regarded as included in this new value, the same as the avoidance of creating a nuisance.

A basic initial observation in this new regime is the existence of clear imbalances between the physical metabolism of our economies and the natural medium in which these economies are located. With regard to the function of the environment as a supplier of production factors, as well as the exhaustion of resources, their exploration and extraction generates a negative environmental impact. Insofar as the environment is seen as a dumping ground for waste generated by economic activity, the rubbish dump and incineration produce unacceptable risks. Other treatment systems, such as composting or the fuel obtained from waste also present similar risks. Consequently, there is also a need for integral management of the complete flow of materials and substances. In the basis of the "new regime" there are the three following principles:

- The prevention ladder. Only materials whose properties are analogous with those of the earth or materials with nutrients that can be used are to be returned to the earth. Sometimes, this principle is also known as "zero emissions".
- The precautionary principle. Society can only authorise activities when there is absolute certainty that they will not be a significant risk for humans and the natural environment.
- The principle of extended producer responsibility. This assigns the responsibility for waste produced by consumption to the economic agents producing the goods and those who participate in transporting or trading with them.

Other important characteristics of the "new regime" are the following:

- The "old regime" principle that every human being has the unlimited right to throw waste away is not accepted. Likewise, authorities are not obliged to accept all the waste deposited or handed over. The obligation to accept is replaced by the right to refuse.
- The geographical scale is not necessarily a local one. Managing the total flow of materials and substances may give rise to cycles and the creation of "cascade effects", which will normally occur on a geographical scale larger than the local one. The possibility is considered of importing and exporting waste (Porter, 2002, chapter 7, or Buclet et al., 2002).
- Priority technologies are those related to prevention and reduction at source: ecodesign, industrial ecology, managing the supply chain, prevention programs, etc. There are also technologies for collecting and separating specific materials, away from the normal waste flows. Tipping and incineration are still present.
- Together with municipal authorities as basic management units, there are regions of waste, where scale economies are attempted. Central and regional governments play a much more important role, since they are the ones who must set up the basic guidelines of public policies and establish certain policy instruments such as bans (for example, banning tips where there is a

heterogeneous flow of waste or prohibiting polystrene containers), taxes and local rates (such as the tax on packaging of consumer products or local rates based on the amount of rubbish generated and collected), other incentives (such as credits and tax exemption for private firms investing in recycling equipment and other investments) or public investment (such as public building of stations where waste-produced energy can be used).

Some Important Events in the Regime Change

At the beginning of the seventies growing affluence in western economies contributed partially to atmospheric degradation and a growth in awareness and knowledge of this degradation. A large number of environmental and ecological groups sprang up, initially in the United States, which introduced the principle of prevention in solid waste policies. In Germany, Holland and Denmark, reacting to lobbying from these groups, a political debate began aimed at applying this principle.

In 1975 the European Union started on the legislative journey by drawing up the Directive 75/442/EEC on waste management (amended in 1991). This directive developed key elements such as the hierarchy of waste and the principles of proximity and self-sufficiency. Moreover, it demanded the introduction of waste management plans.

In 1976 a law came into force in the United States, the Resource Conservation and Recovery Act, aimed at "protecting human health and the environment from the potential hazards of waste elimination, conserving energy and natural resources, reducing the amount of waste generated and ensuring that waste was managed in an environmentally appropriate way".

In 1977 Denmark brought in a law making it compulsory to use returnable containers and banning one-use ones.

At the beginning of the eighties, there were several incidents involving ground and underground water pollution through solid waste tipping. Moreover, there were dioxine emissions from waste incinerator plants. In response stricter environmental standards were introduced, requiring larger scale installation and higher quality standards (which were inevitably costlier), particularly in Germany. The NIMBY syndrome made its appearance with regard to waste tips and incineration plants (it had previously appeared in relation to nuclear power stations). At this time there was also a wave of privatisations in the public sector which also affected the waste sector. In 1988/89 Holland and Germany implemented the prevention principle in their solid waste management plans. This led to selective waste collection, recycling and the introduction of policies regarding packaging.

The 1989 Basle Convention was the starting point for European regulation of international waste transport.

At the beginning of the nineties France introduced the concept of valorisation to consider any method which would enable value to be obtained from waste: incineration with energy recovery, recycling, composting, etc.

In 1991 the Environmental Protection Agency (EPA) in the United States established regulations for landfills. In 1991, when the Töpfer decree came into force, to prevent packaging-produced rubbish, the most important change in Europe regarding waste management and the value of the resources contained in it (Val, 2002) took place. In Holland, the government and industry reached a negotiated agreement on the reduction of packaging waste in 1991. Austria, Belgium, France and Sweden joined this initiative. In 1994, the European Union passed the guideline 94/62/EC "regarding packaging and packaging waste". As Hafkamp (2002) mentions, following on from this guideline, all European Union countries (1) have implemented policies and adapted their legislation to deal with the question of packaging and packaging waste, (2) have made the producers (in the "upstream" sense) responsible for packaging waste after use, (3) have created organisations responsible for collection of industry payments to cover the cost of separating collected waste (only in Germany) and its subsequent recycling, though there are varying definitions of recycling. The French valorisation approach is followed in many countries. The precautionary principle was incorporated.

In 1996 Holland introduced a tipping tax, which was initially 15 euros per ton and later rose to 50 euros a ton. In Italy, the tipping tax was initially between 10 and 15 euros a ton. This tax was also introduced in France. In Holland and in France, the tipping of waste with energy content is prohibited as it is regarded as more suitable for incineration. In 1999 the European Union approved the Directive 99/31/EC "concerning tipping".

MSW Management in the European Union (EU)

As mentioned previously, the waste management strategy of the EU has its origin in the Waste Management Directive of 1975 (75/442/EEC), amended in 1991. The European Union´s approach to waste management is based on three principles:[6]

1. *Waste prevention*: This is a key factor in any waste management strategy. Waste prevention is closely linked with improving manufacturing methods and influencing consumers to demand greener products and less packaging.
2. *Recycling and reuse*: If waste cannot be prevented, as many of the materials as possible should be recovered, preferably by recycling. The European Commission has defined several specific "waste streams" for priority attention, the aim being to reduce their overall environmental impact. These include packaging waste, end of life vehicles, batteries, electrical and electronic waste. EU directives now require Member States to introduce legislation on waste collection, reuse, recycling and disposal of these waste streams.
3. *Improving final disposal and monitoring*: Where possible, waste that cannot be recycled or reused should be safely incinerated, with landfill only used as a last resort. Both these methods need close monitoring because of their potential for causing severe environmental damage.

Packaging and Packaging Waste

In the 1980s and the early 1990s, when implementing new strategies in Europe, emphasis was placed on household packaging waste, since this type of waste revealed the highest rate of increase in industrial countries during this period. Packaging came to account for between 25% and 30% of the total amount of household waste by weight and nearly 50% by volume in

6 *Source*: http://europa.eu.int/comm/environment/waste/index.htm

the early 1990s (Buclet and Godard, 2000b). The European Community first introduced measures on the management of packaging waste with the Directive 85/339/EEC which covered the packaging of liquid beverage containers for human consumption only but it was too vague to bring about the effective harmonization of national policies. On 20 December 1994, the Council of Ministers adopted the 94/62/EC Directive on Packaging and Packaging Waste after three years of negotiations. This Directive[7] aims to harmonise national measures in order to prevent or reduce the impact of packaging waste on the environment and to ensure the functioning of the internal market. It contains provisions on the prevention of packaging waste, on the re-use of packaging and on the recovery and recycling of packaging waste. In accordance with the lastest official data available corresponding to 2003 the recycling rate in the European Union[8] was 55% (glass 60.8%, plastic 26.6%, paper and board 69.1%, metals 60.3%, wood 40.5% and other 6.8%). The rate of recovery or incineration at waste incineration plants with energy recovery was 67.9%.

Landfill of Waste

In accordance with data of 1995 the treatment methods in EU were used in the following proportion: landfilling 67%, incineration 17%, recycling 10%, composting 5%, other 1% (EEA, 1999). Council Directive 99/31/EC on the landfill of waste came into force on 16.07.1999 and the deadline for implementation of the legislation in the Member States was 16.07.2001. The main features of the Directive are the following (Fisher, Petschow and Buclet, 2002):

- Landfill sites are classified into three categories (landfills for hazardous waste, for non-hazardous waste and for inert waste), and the acceptance of some types of waste is prohibited.
- A standardized procedure for the acceptance of waste has been defined.
- A system of operation permits has been established.
- The Member States are in charge of supervising compliance with these waste acceptance procedures.
- A system is prescribed for the control and monitoring of existing landfill sites and a procedure for the closure of landfill sites.
- Conditions are defined to continue the running of an existing landfill site.
- There are some prescriptions calculating a minimum price for landfilling, defining which costs have to be covered by the price.
- Appropriate measures have to be taken to avoid the abandonment, dumping or any uncontrolled disposal of waste.
- Member States are called on to promote compliance with the principles of proximity and self-sufficiency in the case of an intended elimination of waste at the supranational and national level, according to the Waste Management Directive.

---

[7] See http://europa.eu.int/comm/environment/waste/packaging_index.htm

[8] The data correspond to the 15 Member States in 2003 except Greece, Ireland and Portugal. *Source:* http://europa.eu.int/comm/environment/waste/packaging_index.htm (see data 2003).

- In relation to the total volume of biodegradable municipal solid waste generated in 1995, the share of biodegradables destined for landfilling has to be reduced (by weight) to: 75% within five years, 50% within eight years and 35% within fifteen years of the total amount generated.

Waste Incineration[9]

The Directive 2000/76/EC came into force on 29 December 2000. Incorporation into national legislation is necessary by 28 December 2002. From this date on new incinerators will have to comply with the provisions of the Directive. The deadline for existing plants will be 28 December 2005. Then all old Directives (89/369/EEC, 89/429/EEC and 94/67/EC) will be repealed. The aim of the Directive is to prevent or – where that is not practicable- to reduce as far as possible negative effects on the environment caused by the incineration and co-incineration of waste. In particular, it should reduce pollution caused by emissions into the air, soil, surface water and groundwater, and thus lessen the risks which these pose to human health. This is to be achieved through stringent operational conditions and technical requirements and by setting up emission limit values for waste incineration and co-incineration plants within the Community. Considerable reductions will be achieved for acid gases such as nitrogen oxides (NOx), sulphur dioxide (SO_2) and hydrogen chloride (HCl) as well as for heavy metals. Emissions of cadmium throughout the EU are expected to fall from 16 tonnes per year in 1995 to around 1 tonne in 2005. Over the same period, mercury emissions should fall from an annual 36 tonnes to around 7 tonnes. The directive also targets the incineration of non-hazardous waste, which has been identified as the largest ssource of emissions of dioxins and furans into the atmosphere. Incineration was the main treatment method in 1993 in the following countries: Belgium 52%, Denmark 59%, France 44% and Luxembourg 47%.

In Buclet and Godard (2000a) a broad comparative study of municipal waste management in five European countries: France, Germany, Greece, Italy and the Netherlands is presented. In Buclet and Godard (2000b) the following incentive instruments can be found:

- France: Levy on landfilling: 6.15 Ecu/tonne (1998). Average contribution on packaging: 0.0015 Ecu.
- Germany: Contribution per kg. of packaging: between 0.07 Ecu and 1.53 Ecu. Differentiation of 1 to 20 between glass and plastic redistribution of financial support.
- Greece: No instrument. Projects of adaptation of Eco-Emballages (an authorized private French organization, the capital of which is mainly held by fillers (70%), distributors (10%) and material producers (20%), which manage the packaging waste instead of the industries) to the Greek regime.
- Italy: Contribution on plastic bag: 0.07 Ecu/unit. Levy for packaging unit from 0.01 to 0.05 Ecu per packaging.
- The Netherlands: Landfill tax: 15 Ecu / tonne (1996). The Dutch packaging covenant.

---

[9] *Source:* http://europa.eu.int/comm/environment/wasteinc/index.htm

MSW Management in the United States (US)[10]

The Resource Conservation and Recovery Act (RCRA) was enacted by the US Congress in 1976 and amended in 1984. The act´s primary goal is to protect human health and the environment from the potential hazards of waste disposal. In addition, RCRA calls for conservation of energy and natural resources, reduction in waste generated, and environmentally sound waste management practices.

The solid waste management hierarchy ranks the most preferable ways to address solid waste: Source reduction or waste prevention, which includes reuse, is the best approach, followed by recycling/Composting. Waste that cannot be prevented or recycled can be incinerated or landfilled according to proper regulations.

In 2003, US residents, businesses and institutions produced 236.2 million tons of MSW, which is approximately 4.5 pounds of waste per person per day. The total MWS generated in the US has increased by 168.1% from 1960 to 2003 (data from EPA).

The US recycling rates for commonly recycled consumer goods in 2003 are the following: Newspapers 82.4%, Corrugated cardboard boxes 71.3%, steel cans 60.0%, yard trimmings 56.3%, aluminium beer and soft drink cans 43.9%, scrap tyres 35.6%, magazines 33.0%, plastic milk and water bottles 31.9%, plastic soft drink bottles 25.2%, glass containers 22.0% (data from EPA).

The Environmental Protection Agency (EPA) ´s MSW goals for the nation are:

- To recycle 35 percent of MSW generated.
- To reduce waste generation to 4.3 pounds per person per day.
- To empower state, local, and tribal governments to better manage solid waste.
- To provide leadership in source reduction and recycling.
- To build stronger public and private partnerships
- To ensure the environmental soundness of source reduction, recycling, combustion and land disposal.

Repa (2000) points out that in 1997 landfilling was the predominant waste management method in the US, accounting for 69 percent of the waste generated. Recycling accounted for 23 percent, and incineration 8 percent. Repa points out the following trends about US solid waste generation and management:

- The data for the amount of waste being managed off-site were inconsistent and unreliable. The best estimate of the amount of solid waste managed off-site was approximately 437 million tons per year, with tha actual amount likely to be greater.
- The number of US landfills declined significantly since the early 1970s because of landfill closures forced by stricter regulatory programs. The number of landfills declined from an estimated 20,000 in 1970 to 2,800 in 1995. The number of landfills has remained relatively constant since 1995 and is not expected to change significantly.

[10] See: http://www.epa.gov/epaoswer/non-hw/muncpl/facts.htm

- National disposal capacity has increased during the past decade to more than 16 years and probably will increase only when existing capacity is depleted regionally.
- Landfill ownership and the volumes being received at these sites have shifted from the public sector to the private sector. As with the number of landfills, this trend appears to be related to stricter regulations. Whether or not this trend continues cannot be predicted based on the available data.
- Landfill tipping fees increased at a rate of 7 percent annually until 1998, when the fees remained relatively constant at about $32 per ton. This deviation from historical trends is unlikely to continue because many facilities have announced tip fee increases.
- Landfill tip fees historically have been lower than incinerator tip fees, and this trend is not expected to change in the future. In 1982, landfill tip fees were $8.07 per ton and incinerator tip fees were $12.91 per ton, a difference of $4.84. The difference has increased every year since then. In 1998, the average landfill tip fee was $31.81 per ton, and the incinerator tip fee was $58.83 per ton, a difference of $27.02.

MSW Management in Japan

As it is pointed out in Tietenberg (2006), since Japan has a much greater population density than any of the industrialized nations, it has been forced by necessity to face its solid waste problems somewhat earlier than countries with more land area available for disposal. It has done this with a combination of technological solutions, reduction strategies and recycling.

One of the most distinctive features of Japan´s municipal solid waste treatment is the comparatively high rate of incineration, now around 75 percent.

Since the early 1970s, Japanese citizens have been forced to separate combustible from noncombustible trash. Burnable waste, which is some 72% of the total, is burned in some 1,850 incinerators (this quantity can be compared with the 140 large incinerators in the United States). About 80% of the volume of incinerated waste in Japan is used to generate power: What´s left, about 9% of the total waste generated, ends up in landfills.

The recycling rates in Japan in 2004 are the following:[11] tyres 88%, waste paper 69.0%, aluminum cans 86.1%, expanded polystyrene 41.0% and steel cans 87.1%.

The Home Appliance Recycling Law was enacted in April 2001 to promote the reuse of useful resources and to reduce the amount of waste material. This law calls for cooperation among consumers, retailers and manufacturers in recycling unwanted electrical appliances. From April 2001, consumers must apply to the store of purchase for collection of unwanted 1) television sets, 2) air-conditioning units, 3) refrigerators and 4) washing machines. If the store of purchase is far away or unknown, the owner must contact one of the official collection agencies. The owner has to pay a fee to cover the cost of recycling. In the 2002 fiscal year, as many as 10.15 million units of scraped home appliances were collected and recycled under the law.

---

[11] *Source:* http://web-japan.org/stat/19ENV51.html

The Containers and Packaging Recycling Law entered fully into force in April 2000. The aim of this law is to reduce the amount of garbage generated, by promoting more effective use of resources in containers and packaging waste, which accounts for about 60% of houshold waste (by volume). The basic principle of this law is that every player has a role to play in recycling. Consumers should separate theis waste according to category, municipalities should collect the separate waste, and businesses should recycle what has been collected into new products.[12]

The Law for the Promotion of Effective Utilization of Resources entered into force in April 2001. The aim of this law is to promote integrated initiatives for the 3Rs (reduce, reuse, recycle) that are necessary for the formation of a sustainable society based on the 3Rs. In particular, it uses cabinet orders to designate the industries and product categories where businesses are required to undertake 3R initiatives, and stipulates by ministerial ordinances the details of voluntary actions that they should take. Ten industries and 69 product categories have been designated, and actions stipulated include 3R policies at the product manufacturing stage, 3R consideration at the design stage, product identification to facilitate separate waste collection, and the creation of voluntary collection and recycling systems by manufacturers, among other topics.[13]

CONCLUSIONS

In this work a synthesis has been presented of the main economic elements present in the generation and management of municipal solid waste, as well as an analysis of the main policy instruments applied in this context, the present situation in the national and international context and immediate prospects.

In these pages it has been highlighted that, although it has been traditionally regarded as a purely technical problem, municipal solid waste management generates several mechanisms the understanding of which requires a careful economic analysis. Particularly, municipal solid waste produces an external cascade effect, among the different economic agents intervening in the transformation chain, from when the raw materials are extracted, and passing through the phases of production, consumption and generation of waste, till the latter are eliminated. The incompleteness of markets in these phases makes possible the existence of inefficient decisions and, consequently, opens up a field for public sector intervention by means of the application of different economic policies.

These policies can be classified according to the phase in the transformation chain in which they are applied, in *upstream* and *downstream,* or according to their nature, in *command and control* or *incentive* policies, the latter being the ones which fit more fully within economic analysis.

In recent decades we have witnessed a profound change in the principles and practices applied in waste management, a change which has led to a transition from what was called the "old regime" to the "new regime". Among the driving forces of this change an important role has been played by the ever-increasing awareness of the need for responsible management of natural and environmental resources. This need has been made plain in a limitation on the

---

[12] *Source:* http://meti.go.jp/policy/recycle/main/english/law/contain.html

[13] *Source:* http://meti.go.jp/policy/recycle/main/english/law/promotion.html

right to generate and throw away waste, the extension of the scale of regional analysis and the increasing application of public policies based on economic principles.

In the European Union, the United States and Japan the requirements about MSW management have changed in the last decades. Both have incorporated progressively the principles of source reduction and prevention, and the priority of recycling on disposal.

REFERENCES

André, F.J. and Cerdá, E. (2001) "Optimal Sequence of Landfills in Solid Waste Management" *Optimal Control Applications and Methods*, 25: 1-25.

André, F.J. and Cerdá, E. (2004) "Landfill Construction and Capacity Expansion". *Environmental and Resource Economics* 28: 409-434.

André, F.J. and Cerdá, E. (2005) "On the Dynamics of Recycling and Natural Resources" *Environmental and Resource Economics, in press.*

André F.J., Velasco, F. and González, L. (2004) "Intertemporal and Spatial Location of Disposal Facilities". Documento de Trabajo, Centro de Estudios Andaluces E2204/74.

Atri, S. and Schellberg, T. (1995) "Efficient Management of Household Solid Waste: A General Equilibrium Model" *Public Finance Quarterly*, 23: 3-39.

Baumol, W.J. (1977) "On Recycling as a Moot Environmental Issue" *Journal of Environmental Economics and Management*, 4: 83-87.

Beede, D.N. and Bloom, D.E. (1995) "Economics of the Generation and Management of Municipal Solid Waste" National Bureau of Economic Research, Working Paper Series No. 5116.

Bellman, R. (1957) *Dynamic Programming*. Princeton University Press, Princeton, NJ.

Boyer, R. and Orléan, A. (1994) "Persistance et changement des conventions. Deux modèles simples et quelques illustrations", en Orléan, A. (dir.), Analyse économique des conventions, PUF, Paris: 219-247.

Brunner, C.R. (1994) "Incineration". In F. Kreith (ed) *Handbook of Solid Waste*. Mc Graw-Hill.

Bruvoll, A. (1998) "Taxing Virgin Materials: an Approach to Waste Problems" *Resources, Conservation and Recycling*, 22: 15-29.

Buclet, N. and Godard, O. (eds) (2000a) *Municipal Waste in Europe*. A Comparative Study in Building Regimes. Kluwer Academic Publishers, Dordrecht, The Netherlands.

Buclet, N. and Godard, O. (2000b) "Municipal waste management in Europe: a comparison of national regimes", in Buclet, N. and Godard, O. (eds) *Municipal Waste in Europe. A Comparative Study in Building Regimes*. Kluwer Academic Publishers: 203-224.

Buclet, N. *et al.* (2002) "Interactions and main sources of friction between national regimes", in Buclet, N. (ed.) *Municipal Waste Management in Europe. European Policy between Harmonisation and Subsidiarity*. Kluwer Academic Publishers: 27-54.

Calcott, P. and Walls, M. (2000) "Can Downstream Waste Disposal Policies Encourage Upstream *Design for the Environment*" *American Economic Review, Papers and Proceedings* 90: 233-237.

Calcott, P. and Walls, M. (2005) "Waste, Recycling and *Design for the Environment*: Roles for markets and Policy Instruments" *Resource and Energy Economics*, próxima aparición.

Callan, S.J. and Thomas, J.M. (1999) "Adopting a Unit Pricing System for Municipal Solid Waste: Policy and Socio-Economic Determinants" *Environmental and Resource Economics* 14: 503-518.

Chilton, K. (1993) "Solid Waste Policy Should Be Directed by Fundamental Principles, not Ill-founded Feelings" *Resources, Conservation and Recycling*, 8: 1-20.

Choe, C. and Fraser, I. (1999) "An Economic Analysis of Household Waste Management" *Journal of Environmental Economics and Management*, 38: 234-246.

Choe, C. and Fraser, I. (2001) "On the Flexibility of Optimal Policies for Green Design" *Environmental and Resource Economics* 18: 367-371.

Di Vita, G. (2001) "Technological Change, Growth and Waste Recycling". *Energy Economics* 23, 549-567.

Dinan, T.M. (1993) "Economic Efficiency Effects of Alternative Policies for Reducing Waste Disposal" *Journal of Environmental Economics and Management*, 25: 242-256.

Duggal, V.G., Saltzman, C. and Williams, M.L. (1991) "Recycling: and Economic Analysis" *Eastern Economic Journal*, 17: 351-358.

Eea (1999): Environment in the European Union at the turn of the century. Copenhagen.

Eichner, T. and Pethig, R. (2001) "Product Design and Efficient management of Recycling and Waste Treatment" *Journal of Environmental Economics and Management* 41: 109-134.

Eichner, T. and Pethig, R. (2003) "Corrective Taxation for Curbing Pollution and Promoting Green Product Design and Recycling" *Environmental and Resource Economics* 25: 477-500.

Fisher, L., Petschow, U. and Buclet, N. (2002) "The consequences of implementing directives in the national context: the correct answer to the friction between national regimes?", in Buclet, N. (ed.) *Municipal Waste Management in Europe. European Policy between Harmonisation and Subsidiarity*. Kluwer Academic Publishers: 81-110.

Fullerton, D. and Kinnaman, T.C. (1995) "Garbage, Recycling and Illicit Burning or Dumping" *Journal of Environmental Economics and Management*, 29: 78-91.

Fullerton, D. and Kinnaman, T.C. (1996) "Household Responses to Pricing by the Bag" *American Economic Review*, 86: 971-984.

Fullerton, D. and Wu, W. (1998) "Policies for Green Design" *Journal of Environmental Economics and Management*, 36: 131-148.

Gaudet, Moreaux, G., M. and Salant, S.W. (1998) "Intertemporal and Spatial Depletion of Landfills" Fondazione Eni Enrico Mattei Note di Lavoro: 14/98.

Goddard, H.C. (1995) "The Benefits and Costs of Alternative Solid Waste Management Policies" *Resources, Conservation and Recycling*, 13: 183-213.

Hafkamp, W. (2002) "Comparison of National Solid Waste Regimes in Trajectories of Change", en Buclet, N. (ed.) *Municipal Waste Management in Europe. European Policy between Harmonisation and Subsidiarity*. Kluwer Academic Publishers: 7-26.

Halstead, J.M. and PARK, W.M. (1996) "The Role of Economic Analysis in Local Government Decisions: The Case of Solid Waste Management" *Agricultural and Resource Economic Review* 25: 76-82.

Hanley, N., Shogren, J.F. and White, B. (1997) *Environmental Economics. In Theory and Practice*. Macmillan Press LTD, London.

Highfill, J. and Mcasey, M. (1997) "Municipal Waste Management: Recycling and Landfill Space Constraints" *Journal of Urban Economics*, 41: 118-136.

Hoel, M. (1978) "Resource Extraction and Recycling with Environmental Costs" *Journal of Environmental Economics and Management*, 5: 220-235.

Hong, S., Adams, R.M. and Love, H.A. (1993) "An Economic Analysis of Household Recycling of Solid Wastes: The Case of Portland, Oregon" *Journal of Environmental Economics and Management*, 25: 136-146.

Huhtala, A. (1994) Is Environmental Guilt a Driving Force? An Economic Study on Recycling. University of California, Berkeley, PhD. Edited by Lapland University.

Huhtala, A. (1997) "A Post-consumer Waste Management Model for Determining Optimal Levels of Recycling and Landfilling" *Environmental and Resource Economics*, 10: 301-314.

Huhtala, A. (1999) "Optimizing Production Technology Choices: Conventional Production vs. Recycling" *Resource and Energy Economics*, 21: 1-18.

Jacobs, T.L. and Everett, J.W. (1992) "Optimal Scheduling of Consecutive Landfill Operations with Recycling" *Journal of Environmental Engineering*, 118: 420-429.

Jenkins, R.R. (1993) The Economics of Solid Waste Reduction. The Impact of User Fees. Edward Elgar.

Keeler, A.G. and Renkow, M. (1994) "Haul Trash or Haul Ash: Energy Recovery as a Component of Local Solid Waste Management" *Journal of Environmental Economics and Management*, 27: 205-217.

Kiel, K.A. and MCCLAIN, K.T. (1995) "House Prices during Siting Decison Stages: The Case of an Incinerator from Rumor through Operation" *Journal of Environmental Economics and Management*, 28: 241-255.

Kunreuther *et al.* (1987) "A Compensation Mechanism for Siting Noxious Facilites: Theory and Experimental Design" *Journal of Environmental Economics and Management*, 14: 371-383.

Ley, E., Macauley, M.K. and Salant, S.W. (2000) "Restricting the Trash Trade" American Economic Review, *Papers and Proceedings* 90: 243-246.

Ley, E., Macauley, M.K. and Salant, S.W. (2002) "Spatially and Intertemporally Efficient Waste Management: The Costs of Interstate Trade Restrictions" *Journal of Environmental Economics and Management* 43: 188-218.

Lund, J.R. (1990) "Least-Cost Scheduling of Solid Waste Recycling" *Journal of Environmental Engineering*, 116: 182-197.

Lusky, R. (1975a) "Consumers' Preferences and Ecological Consciousness" *International Economic Review,* 16: 188-200.

Lusky, R. (1975b) "Optimal Taxation Policies for Conservation and Recycling" *Journal of Economic Theory*, 11: 315-328.

Morris, G.E. and Holthausen, D.M, Jr. (1994) "The Economics of Household Solid Waste Generation and Disposal" *Journal of Environmental Economics and Management*, 26: 215-234.

Nelson, A.C., Genereaux, J. and Genereaux, M. (1992) "Price Effects of Landfills on House Values" *Land Economics*, 68: 359-365.

Palmer, K., Sigman, H. and Walls, M. (1997) "The Cost of Reducing Municipal Solid Waste" *Journal of Environmental Economics and Management*, 33: 128-150.

Palmer, K. and Walls, M. (1997) "Optimal Policies for Solid Waste Disposal. Taxes, Subsides, and Standards" *Journal of Public Economics*, 65: 193-205.

Pearce, D. and Brisson, I. (1994) "Using Economic Incentives for the Control of Municipal Solid Waste", en Quadrio-Curzio et. al. (eds.) The Management of Muncipal Solid Waste in Europe. Economic, Technological and Environmental Perspectives. Elsevier Science.

Pearce, D.W. and Turner, R.K. (1993) "Market-Based Approaches to Solid Waste Management" *Resources, Conservation and Recycling*, 8: 63-90.

Pearce, D.W. and Turner, R.K. (1995) Economía de los Recursos Naturales y del Medio Ambiente. Colegio de Economistas-Celeste Ediciones, Madrid.

Plourde, C.G. (1972) "A Model of Waste Accumulation and Disposal" Canadian *Journal of Economics*, 5: 119-125.

Porter, R.C. (2002) The Economics of Waste. Resources for the Future, Washington.

Quadrio-Curzio, A., Prosperetti, L. and ZOBOLI, R. (eds) (1994) The Management of Municipal Solid Waste in Europe. Economic, Technological and Environmental Perspectives. Elsevier Science. Developments in Environmental Economics, vol. 5.

Ready, M.J. and Ready, R.C. (1995) "Optimal Pricing of Depletable, Replaceable Resources: The Case of Landfill Tipping Fees" *Journal of Environmental Economics and Management*, 28: 307-323.

Repa, E.W. (2000) "Solid waste disposal trends. A benchmark study on solid waste volumes, the number of landfills, tipping fees and capacity" *Waste Age,* 31, 4: 262-265.

Schulze, W. (1974) "The Optimal Use of Non-Renewable Resources: the Theory of Extraction" *Journal of Environmental Economics and Management,* 1: 53-73.

Shinkuma, T. (2003) "On the Second-Best Policy of Household Waste Recycling" *Environmental and Resource Economics* 24: 77-95.

Sigman H.A. (1995) "A Comparison of Public Policies for Lead Recycling" *The Rand Journal of Economics*, 26: 452-478.

Smith, V.L. (1972) "Dynamics of Waste Accumulation: Disposal Versus Recycling" *Quarterly Journal of Economics*, 86: 600-616.

Tietenberg, T. (2006) *Environmental and Natural Resource Economics.* Seventh Edition. Pearson Eucation, Inc.

Val, A. (2002) *Residuos. El binomio recursos-residuos desde una perspective mediterránea. Análisis de la situación española dentro de la Unión Europea*, en Naredo, J.M. y Parra, F. (eds.) Situación diferencial de los recursos naturales españoles. Economía vs Naturaleza. Fundación César Manrique.

Weinstein, M.C. and Zeckhauser, R.J. (1974) "Use Patterns for Depletable and Recycleable Resources", *The Review of Economic Studies*, «symposium on the economics of exhaustible resources»: 67-88.

Williams, P.T. (1998) *Waste Treatment and Disposal.* John Wiley & Sons, Chichester, England.

Ye, M. and Yezer, A.M.J. (1997) "Where Will We Put the Garbage? Economic Efficiency versus Collective Choice" *Regional Science and Urban Economics*, 27: 47-66.

Zoboli, R. (1994) *The Integrated Use of Economic Instruments in the Policy of Municipal Solid Waste*, en Quadrio-Curzio et. al. (eds.) The Management of Municipal Solid Waste in Europe. Economic, Technological and Environmental Perspectives. Elsevier Science.

In: Solid Waste: New Research
Editors: A. S. Martin and J. D. Suarez

SBN: 978-1-60456-809-7

Chapter 4

INTERSTATE SHIPMENT OF MUNICIPAL SOLID WASTE: 2007 UPDATE*

James E. McCarthy

ABSTRACT

This report, which replaces a 2004 report on the same subject (CRS Report RL32570), provides updated information on interstate shipment of municipal solid waste (MSW). Since the late 1980s, Congress has considered, but not enacted, numerous bills that would allow states to impose restrictions on interstate waste shipments, a step the Constitution prohibits in the absence of congressional authorization. Over this period, there has been a continuing interest in knowing how much waste is being shipped across state lines for disposal, and what states might be affected by proposed legislation. This report provides data useful in addressing these questions. It generally presents data as of 2005.

Total interstate waste shipments continue to rise due to the closure of older local landfills and the consolidation of the waste management industry. More than 42 million tons of municipal solid waste crossed state lines for disposal in 2005, an increase of 8% over 2003. Waste imports have grown significantly since CRS began tracking them in the early 1990s, and now represent 25.3% of the municipal solid waste disposed at landfills and waste combustion facilities. In the last 10 years, reported imports have increased 147%.

Pennsylvania remains the largest waste importer. The state received more than 7.9 million tons of MSW and 1.7 million tons of other non-hazardous waste from out of state in 2005. Most of this waste came from New Jersey and New York. Pennsylvania's waste imports represented 19% of the national total. Virginia and Michigan, the second and third largest importers, received 5.7 million tons and 5.4 million tons from out of state respectively in 2005, each of them about 30% less than the amount received by Pennsylvania.

With the exception of Pennsylvania, each of the 15 largest importers showed an increase in waste imports, compared to our last survey, which provided data as of 2003. Indiana, Michigan, and Wisconsin showed particularly large increases, with Ohio, New York, Oregon, and Georgia also increasing substantially. In each of these states, waste

---

* Excerpted from CRS Report RL34043, dated June 13, 2007.

imports increased by 300,000 tons or more, in some cases substantially more. In all, 30 states had increased imports in the current report, and 11 states reported imports that exceeded 1 million tons.

While waste imports increased overall, Pennsylvania, the leading importer, reported a sharp decline in imports. Pennsylvania's imports fell for the fourth year in a row: about 2.7 million fewer tons of out-of-state MSW were received at Pennsylvania landfills in 2005 than in 2001. Factors causing this decline included the imposition of an additional $4.00 per ton state fee on waste disposal and the absence of rail service at Pennsylvania landfills.

New York remains the largest *exporter* of waste, with New Jersey in second place. Nine other states (Illinois, Missouri, Maryland, Massachusetts, Washington, Minnesota, North Carolina, Indiana, and Florida), the District of Columbia, and the Canadian province of Ontario also exported more than 1 million tons each.

INTRODUCTION

This report provides updated information on interstate shipment of municipal solid waste (MSW). Concerned about increased waste imports, some states have attempted to regulate this commerce, by imposing barriers or requirements specific to waste importation; federal courts, however, have declared such state restrictions unconstitutional. If states are to have such authority, these decisions say, congressional action is required.

Since the late 1980s, Congress has considered, but not enacted, numerous bills that would grant such authority.[1] Over this period, there has been a continuing interest in knowing how much waste is being shipped across state lines for disposal, and what states might be affected by proposed legislation. This report provides data useful in addressing these questions. It updates information provided in earlier CRS reports.[2]

The report presents information gathered through telephone contacts with solid waste officials in the 50 states, the District of Columbia, and the Canadian province of Ontario (which ships large quantities of waste to the United States, principally to Michigan). The data obtained from these contacts are summarized in Tables 1, 2, and 3, and Figures 1 and 2. Table 4 presents additional information, including the names and telephone numbers of state contacts, and in some cases links to detailed reports on solid waste management in the specific state that are available on the Web.

Not all states require reporting of waste imports, and very few track exports, so the available data are incomplete, and in some cases represent estimates rather than actual measurements. In a number of cases, faced with conflicting reports from exporters and importers or no quantitative data at all, the report provides CRS's best estimate, based on discussions with state officials or other sources.

Seven of the states provided data for a period other than calendar year 2005 —either a fiscal year that included part of 2005 or a different calendar year. This adds another layer of imprecision: CRS generally combined data for whatever was the reporting period closest to 2005, even though in these seven cases, this meant combining data from somewhat different time periods. The exceptions are noted in the appropriate tables. As a result, many of the totals reported here represent a best estimate rather than precise figures.

TOTAL SHIPMENTS

The data show that total interstate waste shipments continue to rise:[3] imports in the current survey totaled 42.2 million tons, 17% of the 245.7 million tons of municipal solid waste generated in the United States.[4] Of municipal waste disposed (as opposed to recycled or composted), the percentage is higher. EPA estimates that 79.0 million tons of municipal solid waste were recycled or composted in 2005, leaving 166.7 million tons to be disposed in landfills or incinerators. Of this amount, 25.3% crossed state lines for disposal.[5]

Between CRS's year 2004 report (reporting largely 2003 data) and the current survey (reporting generally 2005 data), imports increased 3.2 million tons, or 8%. Since 1995, reported imports have risen 147%, from 17.1 million tons in 1995 to 42.2 million tons in the current survey.

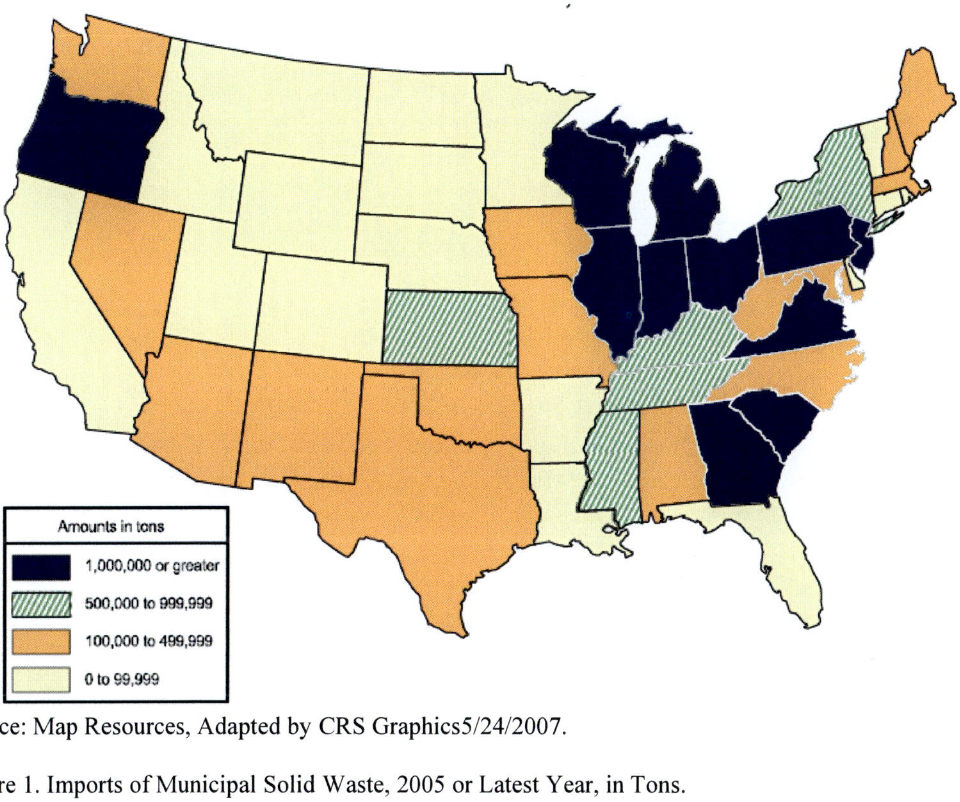

Source: Map Resources, Adapted by CRS Graphics5/24/2007.

Figure 1. Imports of Municipal Solid Waste, 2005 or Latest Year, in Tons.

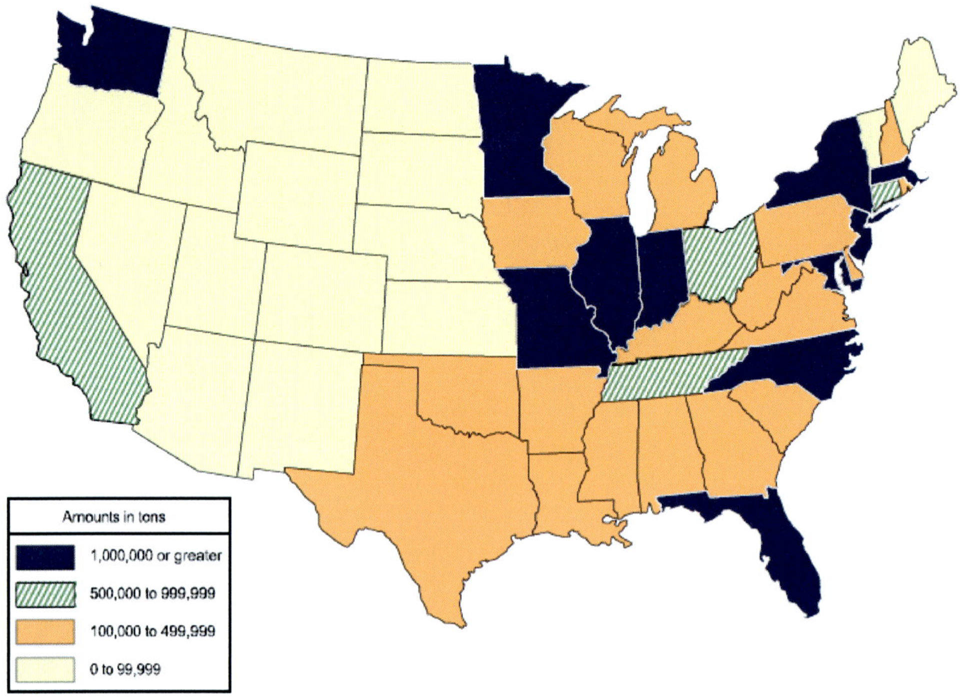

Source: Map Resources, Adapted by CRS Graphics5/24/2007.

Figure 2. Exports of Municipal Solid Waste, 2005 or Latest Year, in Tons.

WASTE IMPORT HIGHLIGHTS

Thirty states had increased imports of municipal waste since 2003, with the largest increases occurring in Indiana, Michigan, and Wisconsin. These three states, along with Pennsylvania and Virginia, accounted for 56% of total municipal waste imports in 2005.

As shown in Table 1, Pennsylvania continues to be the largest waste importer. Disposal facilities in the state received 7.9 million tons of MSW and 1.7 million tons of other nonhazardous waste from out of state in 2005. The amounts represented 39% of all solid waste disposed in the state and 19% of the national total for interstate MSW shipments. Pennsylvania has abundant landfill capacity, relatively low tipping fees, and is near two major states that have a shortage of disposal capacity: New York and New Jersey.

Despite the state's continued predominance on the list of waste importers, Pennsylvania's MSW imports actually declined for the fourth year in a row in 2005 — a cumulative decrease of more than 2.7 million tons. This happened simultaneously with continued growth of interstate waste shipments along the Eastern seabord.

Table 1. Imports of Municipal Solid Waste, 2005 or Latest Year (in tons)

State	Quantity Imported
Pennsylvania	[a] 7,931,984
Virginia	[a] 5,709,441
Michigan	[b,c] 5,442,044
Indiana	[a] 2,428,838
Wisconsin	2,143,133
Illinois	[c] 2,114,898
Oregon	1,795,971
Georgia	1,744,317
New Jersey	1,731,729
Ohio	[a] 1,689,470
South Carolina	[a] 1,243,993
Kansas	800,318
New York	769,083
Tennessee	682,411
Kentucky	663,685
Mississippi	553,772
New Mexico	471,345
Maine	436,412
Arizona	433,400
New Hampshire	402,900
Oklahoma	400,868
Nevada	381,719
Iowa	[d] 300,528
Maryland	[a] 286,011
Texas	259,040
Missouri	227,858
West Virginia	194,917
Massachusetts	169,845
Washington	147,746
Alabama	146,637
North Carolina	[e] 137,298
North Dakota	88,000
Louisiana	e 77,190
California	75,734
Connecticut	43,921
Montana	32,205
Utah	[a] 16,038
Arkansas	7,574
Rhode Island	5,924
Nebraska	d 5,028
South Dakota	1,500
Total	42,194,725

[a] In addition, the state received substantial amounts of

[b] 10/1/2004 - 9/30/2005. industrial, construction and demolition (C and D), or other

[c] Converted from cubic yards by CRS. non-hazardous waste. See Table 4.

[d] 7/1/2004 - 6/30/2005.

[e] 7/1/2005 - 6/30/2006.

Source: CRS, based on data provided by state program officials. See text and Table 4 for qualifications/details.

Several factors appear to have been at work. First, beginning in 2002, Pennsylvania imposed a new state fee of $4.00 per ton on waste disposal. Added to pre-existing fees, the state and local governments in Pennsylvania now collect $7.25 on each ton of waste disposed in the state. This may have provided sufficient economic incentive for some haulers to dispose elsewhere. Second, the state appears to be receiving less waste from New York City, whose Mayor has adopted a goal of shipping all of New York City's waste by rail, rather than truck. Pennsylvania has no landfills served by rail, so some of this waste has been diverted to large landfills in Virginia that do have rail service.

After Pennsylvania, Virginia is the largest waste importer, with imports totaling 5.7 million tons of MSW and 1.3 million tons of other nonhazardous waste. Waste imports to Virginia have increased 45% since 2001, when they totaled 4.1 million tons of MSW and 0.7 million tons of other waste. The state has attempted to restrict imports, but has not been as successful as Pennsylvania, in part because it has chosen a variety of measures that have run afoul of the Constitution's interstate commerce clause. These included a ban on barge shipping of wastes on Virginia rivers, truck regulations that applied only to commercial solid waste transporters, and daily limits on the amount of waste that Virginia landfills could accept.[6]

Michigan, the third-largest waste importer for the past several years, has also seen substantial growth in imports. Significant amounts of waste come to Michigan from Indiana, Illinois, and other neighboring states; but the biggest source, accounting for 69% of Michigan's out-of-state waste, is Ontario, Canada. Ontario is also Michigan's neighbor, but the fact that it lies in a foreign country and that it has large expanses of open land where landfills might be sited seems to have added additional notoriety to its waste shipments. Ontario's shipments to Michigan have grown as Toronto, Canada's largest city, awarded new contracts for waste disposal and closed its last two landfills. At the beginning of 1999, the Toronto area was generating about 2.8 million tons of waste annually, of which about 700,000 tons were shipped to Michigan. By early 2003, however, there was virtually no local disposal capacity in the Toronto area, and almost all of the waste was being shipped to Michigan, where large disposal sites offered very low cost disposal.

In August 2006, the Ontario Minister of the Environment reached an agreement with Michigan's two Senators, under which Ontario will eliminate shipments of municipally managed waste to Michigan by the end of 2010. In return, the Senators agreed not to pursue passage of legislation that would have imposed large inspection fees and other requirements on Ontario's waste shipments to the United States.[7] On September 19, 2006, Toronto's City Council approved a letter of intent to purchase a landfill near London, Ontario, where it is expected to ship its waste as it phases out shipments to Michigan.

The agreement reached by the two Michigan Senators in their exchange of letters with Ontario's Minister of the Environment would not eliminate the majority of the waste shipped from Ontario to Michigan, however. The agreement refers to "municipally managed waste," and specifically uses a 2005 baseline amount of 1.34 million metric tons of municipal waste shipped.[8] About two-thirds of the waste shipped from Ontario is not "municipally managed" – it is waste collected by private haulers and shipped to Michigan landfills under private contracts. These wastes are exported to Michigan either because it provides lower cost disposal options or because the landfills in Michigan are controlled by the same company that collects the waste in Canada. The provincial government and the local governments within the province have no authority to prevent these private waste shipments from leaving Ontario.

(For additional information on Canadian waste import issues, see CRS Report RL33720, *Imports of Canadian Waste.*)

In other highlights from the CRS survey:

- Eleven states reported imports exceeding 1 million tons per year, an increase from 10 in CRS's last survey. Indiana, the additional state, jumped from 11th to 4th on the list with an increase of 1.5 million tons.
- In addition to the 11 states that imported more than a million tons, another 20 states had imports exceeding 100,000 tons.
- Besides the three big increases discussed above (Indiana, Virginia, and Michigan), states that reported major increases in imports compared to CRS's previous survey were Wisconsin, New York, Ohio, Oregon, Georgia, Illinois, Maine, Tennessee, and Kansas, each of which reported an increase of at least 100,000 tons. Growth of waste imports in the Great Lakes states was particularly strong: together, Indiana, Wisconsin, Michigan, Ohio, and Illinois saw an increase of 3.65 million tons in MSW imports.
- New Jersey remains on the list of major importers, with 1.7 million tons of MSW imports in 2005. The state is also a major exporter of waste: receiving states estimated New Jersey's exports at 5.8 million tons. The absence of flow control (local government requirements that waste within their jurisdiction be disposed at local facilities, which were overturned by the courts in the mid-1990s) has led to increased waste exports from New Jersey over the last decade. Waste-to-energy facilities in New Jersey, in turn, began importing MSW in order to replace local waste flowing elsewhere. As a result, large amounts of waste have entered New Jersey from New York in recent years. On April 30, 2007, the Supreme Court held, in the *United Haulers* case, that flow-control ordinances requiring delivery of local waste to a *publicly-owned* processing facility do not violate the Constitution's commerce clause, making it clear that some forms of flow control can survive judicial scrutiny.[9] New Jersey officials do not expect the decision to have much impact on waste exports or imports, however.[10]
- Besides Pennsylvania, only Alabama experienced a major decrease in imports in 2005. Imports to Alabama have been particularly volatile. They declined by almost 270,000 tons (65%) in 2005, compared to 2003, but rebounded 150,000 tons in 2006. Even after that increase, they were less than half the peak amount recorded in 2002.
- Ten other states reported declines in waste imports. The declines were generally small – in half the cases, less than 20,000 tons.
- Although there are no comprehensive data, imports to transfer stations[11] have been a political issue in some locations. Transfer stations are generally located in urban areas and are subject to less stringent regulation than disposal facilities. Heavy truck traffic and odors have aroused concerns in some neighboring communities. Connecticut, Rhode Island, the District of Columbia, and South Carolina have reported significant amounts of out-of-state waste imported to transfer stations, then exported to other states for disposal. A New York City plan to export most of its waste to transfer stations in New Jersey raised substantial controversy, before being rescinded.

Major Exporters

As shown in Table 2, eleven states (New York, New Jersey, Illinois, Missouri, Maryland, Massachusetts, Washington, Minnesota, North Carolina, Indiana, and Florida) and the District of Columbia each exported more than 1 million tons of waste to facilities in other states in the latest reporting period, and 21 other states exported more than 100,000 tons. As noted above, the Canadian province of Ontario also exported a substantial amount of municipal waste (nearly 4 million tons), most of it to Michigan.[12]

Table 2. Exports of Municipal Solid Waste, 2005 or Latest Year (in tons)

State	Quantity Exported
New York	7,198,648
New Jersey	5,772,838
Illinois	4,441,679
Ontario, Canada	[a] 3,976,399
Missouri	2,398,865
Maryland	2,048,204
Massachusetts	1,986,945
Washington	1,745,171
Minnesota	1,085,000
North Carolina	1,074,386
Indiana	1,061,581
District of Columbia	1,061,558
Florida	1,039,611
Ohio	875,005
California	856,509
Connecticut	636,599
Tennessee	518,896
Kentucky	488,157
Texas	460,000
Kansas	446,150
Iowa	409,881
Pennsylvania	338,265
West Virginia	298,238
Wisconsin	263,126
Louisiana	260,588
Alabama	231,700
Virginia	210,688
Mississippi	194,164
New Hampshire	175,000
South Carolina	163,646
Arkansas	161,303
Georgia	125,000
Oklahoma	110,000
Vermont	104,278
British Columbia, Canada	[a] 101,834
Michigan	99,855
Rhode Island	76,077
Maine	71,379
Idaho	63,056
Oregon	52,438

State	Quantity Exported
Delaware	30,000
Alaska	25,201
Nebraska	12,415
Arizona	7,000
Nevada	4,500
North Dakota	3,000
Utah	1,500
Wyoming	200
Total	42,766,533

[a] exports to the United States

Source: CRS, based on data provided by state program officials.

In many cases, the amount is based on data compiled by receiving states. See text and Table 4 entries for additional information and qualifications.

Although the reported amount of total waste exports grew by more than 4 million tons, shipments from the two largest exporting states, New York and New Jersey, did not increase. Compared to CRS's last survey, New York's exports fell more than a million tons to 7.2 million tons in 2005, according to 10 receiving states. New Jersey's estimated exports, 5.8 million tons, remained steady.

By far, the largest growth in exports came from Illinois, whose exports more than doubled, to 4.4 million tons.[13] Most of the exports originate in Cook County (Chicago and its suburbs), which has a relative shortage of disposal capacity. Illinois as a whole has reported a more than doubling of landfill capacity since 1995, but Chicago is located near the border of both Indiana and Wisconsin; so increases in capacity elsewhere in Illinois may not affect disposal decisions in the Chicago metropolitan area.

In all, 10 states and Ontario increased waste exports by more than 100,000 tons each in the period. In addition to Illinois and Ontario, Minnesota and Florida showed the largest increases. Five states and D.C. had decreases of more than 100,000 tons. Besides New York, the others were Georgia, Ohio, Pennsylvania, and Michigan.

Net Imports and Exports

Table 3 combines import and export data to rank the states by net amounts imported or exported. The table shows that 21 states were net importers; 24 plus the District of Columbia were net exporters. Thirty-eight of the 50 states had net imports or exports exceeding 100,000 tons in the reporting period; 22 exceeded 500,000 tons. Perhaps most interesting, given the tendency to identify states as either exporters *or* importers, 25 states both exported *and* imported in excess of 100,000 tons of municipal solid waste, an increase from 23 in CRS's last report.

Table 3. Net Imports/Exports of Municipal Solid Waste, 2005 or Latest Year (in tons)

State	Imports	Exports	Net Imports /Net Exports(-)
Pennsylvania	7,931,984	338,265	7,593,719
Virginia	5,709,441	210,688	5,498,753
Michigan	5,442,044	99,855	5,342,189
Wisconsin	2,143,133	263,126	1,880,007
Oregon	1,795,971	52,438	1,743,533
Georgia	1,744,317	125,000	1,619,317
Indiana	2,428,838	1,061,581	1,367,257
South Carolina	1,243,993	163,646	1,080,347
Ohio	1,689,470	875,005	814,465
New Mexico	471,345	-	471,345
Arizona	433,400	7,000	426,400
Nevada	381,719	4,500	377,219
Maine	436,412	71,379	365,033
Mississippi	553,772	194,164	359,608
Kansas	800,318	446,150	354,168
Oklahoma	400,868	110,000	290,868
New Hampshire	402,900	175,000	227,900
Kentucky	663,685	488,157	175,528
Tennessee	682,411	518,896	163,515
North Dakota	88,000	3,000	85,000
Utah	16,038	1,500	14,538
Nebraska	5,028	12,415	-7,387
Alaska	—	25,201	-25,201
Delaware	—	30,000	-30,000
Idaho	—	63,056	-63,056
Rhode Island	5,924	76,077	-70,153
Alabama	146,637	231,700	-85,063
West Virginia	194,917	298,238	-103,321
Vermont	—	104,278	-104,278
Iowa	300,528	409,881	-109,353
Arkansas	7,574	161,303	-153,729
Louisiana	77,190	260,588	-183,398
Texas	259,040	460,000	-200,960
Connecticut	43,921	636,599	-592,678
California	75,734	856,509	-780,775
North Carolina	137,298	1,074,386	-937,088
Florida	—	1,039,611	-1,039,611
District of Columbia	—	1,061,558	-1,061,558
Minnesota	—	1,085,000	-1,085,000
Washington	147,746	1,745,171	-1,597,425
Maryland	286,011	2,048,204	-1,762,193
Massachusetts	169,845	1,986,945	-1,817,100
Missouri	227,858	2,398,865	-2,171,007
Illinois	2,114,898	4,441,679	-2,326,781
New Jersey	1,731,729	5,772,838	-4,041,109
New York	769,083	7,198,648	-6,429,565

Source: CRS, based on telephone interviews. Data subject to qualifications: see text and Tables 1, 2, and 4.

Table 4. Amount and Destination of Exported MSW, and Amount and Sources of Imported MSW, by State

State	Amount of MSW Exported	Destination of Exported Waste	Amount of MSW Imported	Sources of Imported Waste	Additional Information
Alabama	Receiving states report 231,700 tons of MSW from Alabama in 2005	Tennessee 134,164 tons Mississippi 97,517 tons Georgia 19 tons	146,637 tons in 2005. Imports doubled, to 297,387 tons in 2006, but remained less than half the peak amount (675,000 tons in 2002).	The state does not track the origins of imported waste, but believes it is mostly from Georgia and the Florida panhandle.	Philip Davis, AL Dept. of Environmental Management (334) 271-7755
Alaska	25,201 tons in 2005, according to Washington.	Washington.	No imports.	N.A.	Jennifer Roberts, AK Dept. of Environmental Conservation (907) 269-7553
Arizona	Arizona does not export significant amounts of MSW. There are small flows from border areas to New Mexico, Nevada, and Utah. Based on state estimates, CRS estimates total exports at 7,000 tons.	Arizona estimates that between 1,000 and 10,000 tons may flow to New Mexico; 1,200 tons to Nevada; and 500 tons to Utah.	433,400 tons in 2005.	Nearly all (428,500 tons) from California. Small amounts from Nevada (4,500 tons) and New Mexico (400 tons).	David Janke, AZ Dept. of Environmental Quality (602) 771-4173
Arkansas	Four receiving states reported receiving 161,303 tons from Arkansas, an increase of almost 50,000 tons since 2003.	Missouri ('06) 101,644 tons Mississippi 29,895 tons Texas 22,521 tons Tennessee 7,243 tons	State does not track imports, but believes that imports are relatively small and confined to border areas.	Missouri reported 7,574 tons shipped to Arkansas in 2006.	Susan Speake, AR Dept. of Pollution Control and Ecology (501) 682-0600
British Columbia, Canada	B.C. shipped 101,834 tons to the United States, according to Washington	Washington	N.A.	N.A.	N.A.
California	Receiving states report 856,509 tons of MSW shipped from California. Although exports are substantial, they represent only about 2% of the amount disposed in-state.	Arizona 428,500 tons Nevada 379,009 tons Oregon 49,000 tons	75,734 tons in 2005.	State does not keep track of where waste comes from.	Sherry Sala-Moore, CA Integrated Waste Management Board (916) 341-6204 www.ciwmb.ca.gov/ lgcentral/drs/Reports/ Statewide/SWTotals. Asp

Table 4. (Continued).

State	Amount of MSW Exported	Destination of Exported Waste	Amount of MSW Imported	Sources of Imported Waste	Additional Information
Colorado	State does not track exports. Very small amounts may be exported to neighboring states.	Kansas, Nebraska, New Mexico	State does not track imports. Small amounts may be imported from Kansas and Nebraska.	Kansas, Nebraska	Charles Johnson, CO Dept. of Public Health and Environment (303) 692-3348
Connecticut	Six states reported receiving 636,599 tons from Connecticut in 2005.	New York 218,013 tons Pennsylvania 201,700 tons Ohio 131,801 tons Massachusetts 81,151 tons Georgia 3,869 tons Michigan 36 tons West Va. 29 tons	Connecticut reports 43,921 tons of MSW imports in 2005.	Mass. 36,924 tons New York 3,769 tons Rhode Isl. 3,218 tons	Judy Belaval, CT Dept. of Environmental Protection (860) 424-3237
Delaware	The state does not track MSW exports. CRS estimates exports at 30,000 tons in 2005 based on reports from receiving states, a decline of about 75% since 2003.	Virginia 18,537 tons Penn. 8,741 tons Small amounts to Maryland and New Jersey.	The state does not track MSW imports but says it is likely a negligible amount. All MSW landfills in the state are owned by the state and are prohibited from accepting out-of-state waste.	N.A.	Nancy Markur, DE Dept. of Natural Resources and Environmental Control, (302) 739-9403

Table 4. (Continued).

State	Amount of MSW Exported	Destination of Exported Waste	Amount of MSW Imported	Sources of Imported Waste	Additional Information
District of Columbia	Receiving states reported receiving at least 1,061,558 tons in 2005, the bulk of which went to Virginia.	Virginia 1,059,700 tons Penn. 1,858 tons An uncertain amount went to Maryland, as well.	There are no disposal facilities in the District of Columbia, but D.C. has imported substantialamounts of waste from Maryland to transfer stations located in the District. This waste is then exported for disposal. According to D.C., about one quarter of the waste handled at D.C. transfer stations originates in Maryland.	Maryland.	Thomas Henderson, D.C. Dept. of Public Works, Solid Waste Division, (202) 645-141
Florida	The state does not track exports. Georgia reports receiving over 1 million tons of MSW from Florida in 2005. Exports to Georgia increased 350,000 tons since 2003, but still represent only 3% of Florida's waste generation.	Georgia 1,039,611 tons Small amounts may go to Alabama.	The state does not track imports. There is little incentive to import, since disposal is less expensive in Georgia, and there are no major out-of-state cities near the Florida border.	N.A.	Peter Goren, FL Dept. of Environmental Protection (850) 245-8714

Table 4. (Continued).

State	Amount of MSW Exported	Destination of Exported Waste	Amount of MSW Imported	Sources of Imported Waste	Additional Information
Georgia	CRS estimates 125,000 tons of exports based on information available from three receiving states. Exports decreased from an estimated 600,000 tons in 2003.	Alabama 75,000 tons S. Carolina 28,810 tons Tennessee 17,056 tons	1,744,317 tons in 2005. Waste imports have increased by 750,000 tons since 2002.	Florida 1,039,611 tons N. Jersey 394,747 tons S. Carolina 81,738 tons New York 75,345 tons N. Carolina 42,668 tonsRh. Island 38,687 tons Tennessee 30,083 tons Maryland 29,454 tons 12 others 11,984 tons	Scott Henson, GA Dept. of Natural Resources (404) 362-4533
Hawaii	No exports of MSW in 2005. Proposals to export waste from Oahu to Washington state or Idaho are under consideration.	N.A.	No imports of MSW.	N.A.	Gary Siu, HI Dept. of Health (808) 586-4244
Idaho	Idaho does not track exports. Three receiving states report 63,056 tons in 2005.	Washington 32,256 tons Montana 29,000 tons Oregon 1,800 tons	Idaho does not track imports, but says there is not a large amount of waste imported currently. Idaho Waste Systems has applied for permission to import substantial quantities from Hawaii, however.	Small amounts from Oregon and Nevada.	Dean Ehlert, ID Dept. of Environmental Quality (208) 373-0416

Table 4. (Continued).

State	Amount of MSW Exported	Destination of Exported Waste	Amount of MSW Imported	Sources of Imported Waste	Additional Information
Illinois	Six neighboring states report receiving 4,441,679 tons of MSW from Illinois in 2005. Exports more than doubled since 2003.	Indiana 2,522,635 tons Wisconsin 1,412,153 tons Michigan 416,538 tons Missouri 71,095 tons Iowa 12,926 tons Kentucky 6,332 tons	The state reports 2,114,898 tons of imports in 2005. (Data converted from cubic yards to tons by CRS.)	Missouri (76%) Iowa (19%) Indiana (3%) Wisconsin (2%) Small amounts from 6 other states.	Ellen Robinson, IL Environmental Protection Agency (217) 782-9288 http://www.epa.state.il.us/l and/landfill-capacity/index .html
Indiana	Five receiving states reported a total of 1,061,581 tons of MSW from Indiana in 2005.	Michigan 731,270 tons Kentucky 170,870 tons Ohio 97,518 tons Illinois 61,854 tons Virginia 69 tons	2,428,838 tons of MSW in 2005, an increase of 1.5 million tons from 2003. The state also received 658,000 tons of other solid waste from out of state in 2005.	Illinois 2,122,945 tons Ohio 115,489 tons Kentucky 109,786 tons Michigan 65,521 tons 23 others 15,097 tons	Michelle Weddle, IN Dept of Environmental Management (317) 233-4624 www.in.gov/idem/catalog/ documents/land/far05.pdf
Iowa	409,881 tons in 2005.	Illinois 398,112 tons Missouri 6,704 tons Nebraska 5,028 tons Wisconsin 37 tons (Exports to Nebraska do not include waste directly hauled without passing through a transfer station.)	The state reported a total of 300,528 tons in FY2005. Imports declined to 281,925 tons in FY2006.	Minnesota 265,939 tons Illinois 11,874 tons Missouri 10,857 tons Nebraska 8,952 tons Wisconsin 2,901 tons	Mark Warren, IA Dept of Natural Resources (515) 281-4968
Kansas	Kansas reports MSW exports of 446,150 tons in 2005. Waste exports “went way down” in 2006, because a new landfill opened in Kansas.	Oklahoma 400,868 tons Missouri 45,282 tons	800,318 tons of MSW in 2005, almost all from Missouri.	Missouri 769,356 tons Oklahoma 27,499 tons Nebraska 3,463 tons	Christine Mennicke, KS Dept. of Health and Environment (785) 296-0724

Table 4. (Continued).

State	Amount of MSW Exported	Destination of Exported Waste	Amount of MSW Imported	Sources of Imported Waste	Additional Information
Kentucky	488,157 tons in 2005, a 48% increase since 2003.	Tenn. 283,836 tons Indiana 141,365 tons Ohio 58,679 tons Illinois 4,277 tons	663,685 tons in 2005. Imports in 2006 rose slightly to 686,151 tons.	Ohio 249,902 tons Indiana 170,870 Tenn. 126,416 West Va 106,936 Smaller amounts from Illinois, Missouri, New Jersey, and Virginia.	Allan Bryant, KY Dept. for Environmental Protection (502) 564-6716
Louisiana	Neighboring states reported 260,588 tons in 2005. Little change from 2003.	Texas 152,615 tons Mississippi 107,973 tons	77,190 tons in FY2006 (July 1, 2005-June 30, 2006).	Mississippi 60,000 tons Texas 10,300 tons Arkansas 6,500 tons (CRS estimates based on La. data.)	John Rogers, LA Dept. of Environmental Quality (225) 219-3266
Maine	Maine reports exports of 71,379 tons in 2005.	About 15,000 tons went to New Brunswick, Canada, and the rest to New Hampshire.	Maine imported 436,412 tons of MSW and C and D waste in 2005.	Facilities don't report state of origin, but 2/3 to 3/4 of the waste is believed to come from Massachusetts. The rest probably comes from New Hampshire.	George MacDonald, ME Dept of Environmental Protection (207) 287-5759
Maryland	Receiving states reported receiving 2,048,204 tons from Maryland in 2005. 97% of the exports wentto Virginia.	Virginia 1,992,313 tons Georgia 29,454 tons Penn. 26,350 tons West Virginia 87 tons	The state reported receiving 286,011 tons of out-of-state MSW, and 245,835 tons of other waste, mostly C and D in 2005. Imports increased 37% compared to calendar year 2004.	Massachusetts, New York, West Virginia, Pennsylvania, and D.C.	Edward Dexter, MD Dept of the Environment (410) 537-3318 http://www.mde.state.md.us/assets/document/SW_Managed_in_MD_Report_CY_2005.pdf

Table 4. (Continued).

State	Amount of MSW Exported	Destination of Exported Waste	Amount of MSW Imported	Sources of Imported Waste	Additional Information
Massachusetts	Receiving states reported a total of 1,986,945 tons from Massachusetts in 2005.	So. Carolina 475,495 tons Georgia 394,747 tons Maine 300,000 tons N. Hampshire 281,375 tons New York 216,661 tons Ohio 168,740 tons Maryland 101,367 tons Connecticut 36,924 tons Rhode Island 5,924 tons Pennsylvania 5,417 tons Michigan 273 tons Virginia 22 tons	In 2005, Massachusetts reported importing a total of 169,845 tons.	Connecticut 81,151 tons N. Hampshire 41,079 tons Rhode Island 30,534 tons Vermont 16,391 tons New York 627 tons Maine 63 tons	Brian Holdridge, MA Dept. of Environmental Protection (617) 292-5578
Michigan	The state does not track exports, but three neighboring states reported 99,855 tons from Michigan in 2005, a decrease of 125,000 tons since 2003.	Indiana 65,521 tons Ohio 32,658 tons Wisconsin 1,676 tons	In FY2005 (10/04 - 9/05), imports of MSW were 5,442,044 tons, an increase of almost half a million tons since FY 2003. Michigan also imported 721,000 tons of industrial solid waste. (Data converted from cubic yards to tons by CRS.) Imports leveled off in FY2006, increasing less than 1%.	Ontario 3,781,171 tons Indiana 731,270 tons Illinois 416,538 tons Ohio 299,791 tons Wisconsin 211,648 tons Three other states (New York, Massachusetts, and New Jersey) shipped small amounts.	Christina Miller, MI Dept. of Environmental Quality (517) 373-4741 http://www.deq.state.mi.us /documents/deq-whm-stsw – ReportSolidWasteLandfilledFY2005.pdf
Minnesota	In 2005, the state exported about 1,085,000 tons.	Wisconsin 729,264 tons Iowa 265,939 tons No. Dakota 88,000 tons So. Dakota 1,500 tons	According to the state, a negligible amount has been imported.	N.A.	Jim Chiles, MN Pollution Control Agency (651) 296-7273

Table 4. (Continued).

State	Amount of MSW Exported	Destination of Exported Waste	Amount of MSW Imported	Sources of Imported Waste	Additional Information
Mississippi	194,164 tons, according to receiving states.	Tennessee 134,164 tons Louisiana 60,000 tons (FY06)	553,772 tons in 2005. Imported amounts have been relatively stable since 2002.	Tennessee 318,391 tons Louisiana 107,973 tons Alabama 97,517 tons Arkansas 29,895 tons	Pradip Bhowal, MS Dept. of Environmental Quality (601) 961-5082 http://www.deq.state.ms.us/MDEQ.nsf/pdf/SW_2005StatusReport/$File/AnnualReport2005%20-%20Web%20Version.pdf?OpenELement
Missouri	2,398,865 tons in 2005; 2,520,071 tons in 2006.	Illinois 1,598,625 tons Kansas 769,356 tons Iowa 10,857 tons Tennessee 9,723 tons Arkansas ('06) 7,574 tons Kentucky 2,730 tons	227,858 tons in 2006, a slight increase over 2003.	Arkansas 101,644 tons Illinois 81,917 tons Kansas 37,594 tons Iowa 6,704 tons	Glenda Marshall-Griffin, MO Dept. of Natural Resources (573) 526-3843
Montana	Montana does not track exports, and is not believed to export any significant amount of MSW.	N.A.	32,205 tons in 2005 — almost identical to the amount in 2003.	Idaho 29,000 tons No. Dakota 3,000 tons The rest from Wyoming and Utah.	Pat Crowley, MT Dept. of Environmental Quality (406) 444-5294
Nebraska	The state does not collect records on MSW exports, but Iowa and Kansas reported receiving 12,415 tons from Nebraska in 2005. Iowa alone received 23,628 tons from Nebraska in FY 2006.	Iowa 8,952 tons Kansas 3,463 tons	The state does not collect records on MSW imports. Iowa reports sending Nebraska 5,028 tons of MSW in FY2005.	Iowa.	Keith Powell, NE Dept. of Environmental Quality (402) 471-4210

Table 4. (Continued).

State	Amount of MSW Exported	Destination of Exported Waste	Amount of MSW Imported	Sources of Imported Waste	Additional Information
Nevada	Arizona estimates that it received 4,500 tons of MSW from Nevada. In addition, a small amount is exported to Idaho from border communities in the northeast corner of the state.	Arizona, Idaho.	381,719 tons in 2005.	Almost all (379,009 tons) from California. A small amount is imported from neighboring communities in Utah and Arizona.	Dave Simpson, NV Division of Environmental Protection (775) 687-9469
New Hampshire	CRS estimates exports of 175,000 tons in 2005, based on reports from receiving states.	Mostly to Maine; 41,000 tons to Massachusetts.	In 2005, New Hampshire imported 402,900 tons of MSW, primarily from Massachusetts. Imports were unchanged compared to 2002.	Mass. 281,375 tons Maine 54,000 tons Vermont 49,800 tons Conn. 10,661 tons Rhode Island 6,856 tons	Donald Maurer, NH Dept. of Environmental Services (603) 271-3713
New Jersey	5,772,838 tons in 2005, according to eight receiving states.	Penn. 4,512,908 tons Georgia 394,747 tons Virginia 334,009 tons Ohio 316,656 tons S. Carolina 155,716 tons New York 56,136 tons West Va. 2,086 tons Maryland 580 tons	1,731,729 tons in 2005, 94% from New York.	New York 1,639,916 tons Penn. 70,950 tons Internat'l 16,689 tons 9 other states 4,174 tons	Ray Worob, NJ Dept. of Environmental Protection (609) 984-6903
New Mexico	Texas and Arizona report receiving small amounts of waste from New Mexico.	Texas and Arizona.	471,345 tons were imported in 2005, a decrease of about 65,000 tons since 2003.	Texas 450,000 tons Colorado 17,000 tons The rest is from Arizona, Oklahoma, Mexico, and possibly Utah.	Connie Pasteris, NM Environment Dept. (505) 476-3561

Table 4. (Continued).

State	Amount of MSW Exported	Destination of Exported Waste	Amount of MSW Imported	Sources of Imported Waste	Additional Information
New York	Ten importing states report a total of 7,198,648 tons from New York in 2005, a decrease of over 1 million tons since 2003. New York facilities reported exports of 4,070,503 tons in 2005.	Penn. 3,075,953 tons Virginia 1,803,754 tons New Jersey 1,639,916 tons Ohio 583,999 tons Georgia 75,345 tons West Va. 13,810 tons Connecticut 3,769 tons Michigan 1,325 tons Massachusetts 627 tons Kentucky 150 tons	New York reports that 769,083 tons of MSW were imported in 2005, an increase of 450,000 tons since 2003. The state also imported 390,000 tons of other solid waste in 2005.	Conn. 218,013 tons Mass. 216,661 tons Ontario 195,228 tons New Jersey 56,136 tons Pennsylvania 41,368 tons Vermont 38,087 tons Quebec 2,114 tons N. Hampshire 1,476 tons	Gerard Wagner, NY State Dept. of Environmental Conservation (518) 402-8662
North Carolina	1,074,386 tons in 2005, according to receiving states. In addition, the state exported 96,001 tons to a South Carolina transfer station, which, after baling, were sent back to North Carolina for disposal. Exports account for slightly over 10% of the waste generated in the state.	S. Carolina 554,074 tons Virginia 418,868 tons Tennessee 56,806 tons Georgia 42,668 tons West Va. 1,970 tons	137,298 tons in FY2006 (July 2005-June 2006). Does not include 107,888 tons of waste imported from a South Carolina transfer station, which originally received the waste from North Carolina.	S. Carolina 80,661 tons Virginia 56,637 tons	Ellen Lorscheider, NC Dept. of Environment and Natural Resources (919) 508-8499 http://wastenot.enr.state.nc.us/swhome/AR05-06.pdf
North Dakota	Montana estimates that North Dakota exported 3,000 tons to Montana in 2005.	Montana	88,000 tons in 2005, according to Minnesota.	Minnesota	Steve Tillotson, ND Dept. of Health (701) 328-5166

Table 4. (Continued).

State	Amount of MSW Exported	Destination of Exported Waste	Amount of MSW Imported	Sources of Imported Waste	Additional Information
Ohio	857,005 tons in 2005, a decrease of almost 250,000 tons since 2003.	Michigan 299,791 tons Kentucky 249,902 tons West Va. 161,024 tons Indiana 115,489 tons Pennsylvania 29,832 tons Geogia 815 tons Virginia 152 tons	Ohio imported 3,024,452 tons of solid waste in 2005, but 43% of it was C and D waste, industria waste, and other non-MSW. Imports of general solid waste, the equivalent of MSW, totalled 1,689,470 tons.	Ohio imported waste from 27 states. The largest source were New York (35%), New Jersey (19%), Pennsylvania (13%), Massachusetts (10%), Connecticut (8%), Indiana (6%), West Virginia (4%), and Kentucky (3%).	Michelle Kenton, OH Environmental Protection Agency (614) 728-5368 http://www.epa.state.oh.us /pic/facts/2005_out_of_sta te_waste.pdf
Oklahoma	CRS estimates exports at 110,000 tons in 2005, based on reports from receiving states.	Texas about 80,000 tons Kansas 27,499 tons Small amounts to New Mexico.	State does not track imports. Kansas reports that 400,868 tons of waste were shipped from the Wichita area to Oklahoma in 2005, but the quantity imported dropped significantly in mid to late-2006, when a new landfill opened in Kansas.	Mostly from Kansas.	John Roberts, OK Dept. of Environmental Quality(405) 702-5100
Ontario, Canada	Ontario shipped 3,976,399 tons of MSW to the United States in 2005, according to receiving states. Michigan received 95% of the total. (Data for Michigan are for FY2005 and were converted from cubic yards to tons by CRS.)	Michigan 3,781,171 tons New York 195,228 tons	None.	N.A.	Bruce Pope, Ontario Ministry of Environment and Energy (416) 325-4420

Table 4. (Continued).

State	Amount of MSW Exported	Destination of Exported Waste	Amount of MSW Imported	Sources of Imported Waste	Additional Information
Oregon	Oregon exported 52,438 tons of MSW in 2005.	Washington and Idaho.	Oregon imported 1,795,971 tons of MSW in 2005. Imports accounted for 37% of all the waste disposed in Oregon that year.	Wash. 1,745,171 tons California 49,000 tons Idaho 1,800 tons	Judy Henderson, OR Dept. of Environmental Quality (503) 229-5521
Pennsylvania	The state does not track exports. According to neighboring states, Pennsylvania exported 338,265 tons of MSW in 2005.	Ohio 214,951 tons New Jersey 70,950 tons New York 41,368 tons West Virginia 9,513 tons Virginia 1,483 tons (Exports to Ohio estimated by CRS, based on Ohio data.)	7,931,984 tons in 2005, a decline of 2.7 million tons since 2001. The state is still, by far, the largest importer of MSW, representing about 20% of the national total of imports. In addition to MSW, Pennsylvania received 1.7 million tons of other solid waste from out of state in 2005.	New Jersey and New York accounted for nearly 96% of Pennsylvania's MSW imports in 2005. NJ 4,512,908 tons NY 3,075,953 tons Conn. 201,700 tons West Va. 68,264 tons Ohio 29,832 tons Maryland 26,350 tons 6 others 16,976 tons	Sally Lohman, PA Dept. of Environmental Protection (717) 787-7382 [http://www.depweb.state.pa.us/landrecwaste/cwp/view.asp?a=1238 and Q=464453 and PM=1]
Rhode Island	Receiving states reported 76,077 tons of MSW from Rhode Island in 2005.	Georgia 38,687 tons Mass. 30,534 tons N. Hampshire 6,856 tons Small amounts to Connecticut and New Jersey.	Massachusetts reports sending 5,924 tons of MSW to RI. Officially, however, RI does not accept MSW from out-ofstate. In 2005, all MSW imported to RI was reported as sent back outof-state for disposal.	Massachusetts	Robert Schmidt, RI Dept. of Environmental Management (401) 222-2797 x7260

Table 4. (Continued).

State	Amount of MSW Exported	Destination of Exported Waste	Amount of MSW Imported	Sources of Imported Waste	Additional Information
South Carolina	Receiving states reported 163,646 tons of waste from South Carolina.	Georgia 81,738 tons NC (FY06) 80,661 tons West Va. 748 tons Virginia 499 tons	South Carolina imported 1,243,993 tons of MSW in FY2005 (7/04-6/05), plus 284,106 tons of other solid waste disposed at MSW landfills.	No. Carolina 554,074 tons Mass. 475,495 tons New Jersey 155,716 tons Texas 29,882 tons Georgia 28,810 tons Non-MSW came mostly from Georgia, Delaware, and North Carolina.	Pete Stevens, SC Dept. of Health and Environmental Control (803) 896-4149 http://www.scdhec.gov/recycle/forms/msw05f.pdf
South Dakota	The state does not track exports of MSW.	N.A.	The state does not track imports of MSW. Minnesota reports having shipped 1,500 tons of waste to South Dakota in 2005.	Minnesota	Jim Wente, SD Dept. of Environment and Natural Resources (605) 773-3153
Tennessee	Six neighboring states report receiving 518,896 tons of waste from Tennessee in 2005, an increase of about 70% since 2001.	Mississippi 318,391 tons Kentucky 126,416 tons Virginia 39,805 tons Georgia 30,083 tons The remainder went to Indiana and W. Virginia.	682,411 tons in 2005, 741,560 tons in 2006. Imports increased 28% from 2003 to 2006.	Kentucky 283,836 tons Virginia 147,485 Miss. 134,164 NC 56,806 The remainder came from 5 other states. (2005 data)	A. Wayne Brashear, TN Dept. of Environment and Conservation (615) 532-8010
Texas	460,000 tons.	New Mexico 450,000 tons Louisiana 10,300 tons (FY06)	259,040 tons in 2005.	Louisiana 152,615 tons Oklahoma 83,219 tons Arkansas 22,521 tons Small amounts from New Mexico and Kansas. (Oklahoma and Arkansas are estimated based on Texas data.)	Edward Block, TX Commission on Environmental Quality (512) 239-6613

Table 4. (Continued).

State	Amount of MSW Exported	Destination of Exported Waste	Amount of MSW Imported	Sources of Imported Waste	Additional Information
Utah	CRS estimates exports at 1,500 tons. As in previous years, about 1,000 tons of waste went from Wendover, Utah, to Wendover, Nevada. Also, Arizona reports about 500 tons of waste from Utah. Perhaps 50 tons to Montana.	Nevada, Arizona, Montana	16,038 tons of MSW in 2005, plus 275,837 tons of industrial waste.	Arizona	Ralph Bohn, UT Dept. of Environmental Quality(801) 538-6794
Virginia	The state does not track MSW exports. Six states report 210,688 tons of exports from Virginia.	Tennessee 147,485 tons NC (FY06) 56,637 tons West Va. 5,321 tons Penn. 918 tons Kentucky 283 tons Georgia 44 tons	Virginia remains the second-largest waste importer. The state imported 5,709,441 tons of MSW in 2005 and 1.3 million tons of other waste (mostly C and D waste, incinerator ash, and industrial waste). Imports increased by about 400,000 tons compared with 2003.	Maryland 1,992,313 tons New York 1,803,754 tons DC 1,059,700 tons N. Carolina 418,868 tons New Jersey 334,009 tons Smaller amounts from 13 other states.	Kathy Frahm, VA Dept. of Environmental Quality (804) 698-4376 http://www.deq.virginia.gov/waste/aswrs.html
Washington	1,745,171 tons of MSW in 2005, according to Oregon. Washington has over 200 million tons of disposal capacity (38 years at current disposal rates), but because of contractual arrangements, the state exports substantial amounts of waste.	Oregon.	147,746 tons of MSW in 2005, plus 67,112 tons of other waste.	B.C., Canada 101,834 tons Oregon 45,554 tons Idaho 32,256 tons Alaska 25,201 tons Montana 13 tons	Ellen Caywood, WA Dept. of Ecology (360) 407-6132 http://www.ecy.wa.gov/pubs/0607024.pdf

Table 4. (Continued).

State	Amount of MSW Exported	Destination of Exported Waste	Amount of MSW Imported	Sources of Imported Waste	Additional Information
West Virginia	No tracking system. Eight receiving states reported 298,238 tons of waste from West Virginia. Exports virtually unchanged since 2003.	Kentucky 106,936 tons Ohio 74,301 tons Pennsylvania 68,264 tons Virginia 38,114 tons Maryland 8,844 tons 3 other states 1,779 tons (Exports to Ohio estimated by CRS, based on Ohio data.)	194,917 tons in 2005, a decrease of almost 30% since 2003. Imports represented about 10% of total waste disposal in West Virginia in 005.	Ohio 161,024 tons NY 13,810 tons Penn. 9,513 tons Virginia 5,623 tons The rest from 6 other states.	Jan Borowski, WV Solid Waste Management Board (304) 926-0448
Wisconsin	The state does not collect export data, but four receiving states reported 263,126 tons of Wisconsin exports in 2005, an increase of 23% since 2003.	Michigan 211,648 tons Illinois 47,056 tons Iowa 2,901 tons Indiana 1,521 tons	2,143,133 tons in 2005, an increase of 77% since 2003. Imports from Illinois and Minnesota both increased substantially.	Illinois 1,412,153 tons Minn. 729,264 tons Michigan 1,676 tons	Lindsey Miller, WI Dept. of Natural Resources (608)266-2111
Wyoming	The state does not collect export data. Montana reported about 200 tons from Wyoming.	Montana	The state does not collect import data. A few tons a day may enter the state.	N.A.	Bob Doctor, WY Dept. of Environmental Quality (307) 473-3468

N.A. = not available

Source: CRS, based on information provided by state program officials.

Several factors are at work here. In the larger states, there are sometimes differences in available disposal capacity in different regions within the state. Areas without capacity may be closer to landfills (or may at least find cheaper disposal options) in other states. A good example is Illinois: the Chicago area, which is close to two other states, exports significant amounts of waste out of state. Downstate, however, Illinois has substantial available landfill capacity, and imported 2 million tons from St. Louis, other locations in Missouri, and Iowa.

As noted earlier, the movement of waste also represents the regionalization and consolidation of the waste industry. In 2005, the three largest firms (Waste Management, Allied Waste, and Republic Services) accounted for 66% of total revenues of the industry's 100 largest firms.[14] These large firms offer integrated waste services, from collection to transfer station to disposal site, in many locations. Often, they ship waste to their own disposal facility across a border, rather than dispose of it at an in-state facility owned by a rival. As small landfills continue to close — the number of U.S. landfills declined 63% between 1993 and 2004, from 4,482 to 1,654[15] — this trend toward regionalization, consolidation, and waste shipment across state lines is likely to continue.

Additional Information

The remainder of this report consists of a table summarizing waste import and export data, by state. All 50 states and the District of Columbia are listed in alphabetical order, with data for the amount of waste exported, destination of exports, amount of waste imported, source of imports, and a state agency contact for additional information.

References

[1] Legislation on interstate shipment of waste has been introduced in every Congress since the 100th. In the 104th Congress, the Senate passed S. 534, which would have granted states authority to restrict new shipments of municipal solid waste from out of state, if requested by an affected local government. In the 103rd Congress, both the House and Senate passed interstate waste legislation (H.R. 4779 and S. 2345), but lack of agreement on common language prevented enactment. For a discussion of the issues addressed in these bills, see CRS Report RS20106, *Interstate Waste Transport: Legislative Issues*, by James E. McCarthy.

[2] This report replaces CRS Report RL32570, *Interstate Shipment of Municipal Solid Waste: 2004 Update*, by James E. McCarthy. Earlier reports, many of which are now out of print but available directly from the author, were prepared in 2002, 2001, 2000, 1998, 1997, 1996, 1995, and 1993.

[3] We rely on imports rather than exports as our measure of total shipments, because we believe that waste management facilities and states have a greater interest in accurately measuring imports than they do exports. Often the amounts received and their source are subject to formal legal reporting requirements and/or fees, with penalties for failure to report. Exports are not generally subject to such requirements.

[4] Because many of the larger importing states now differentiate MSW from other nonhazardous waste imports, we compared total MSW imports to EPA's national estimate of MSW generation (245.7 million tons in 2005). For EPA data on waste generation, see *Municipal Solid Waste in the United States: 2005 Facts and Figures*, at [http://www.epa.gov/msw/pubs/mswchar05.pdf]. State-reported waste generation, summarized in *BioCycle* magazine's biannual survey, is substantially higher (509 million tons in 2004) but may include other nonhazardous waste, provided it was disposed at MSW facilities. For state-reported data, see Phil Simmons, Nora Goldstein, Scott M. Kaufman, Nickolas J. Themelis, and James Thompson, Jr., "The State of Garbage in America," *BioCycle*, April 2006, p. 26. Removing Canadian waste from the total imports would also reduce the percentage of waste crossing state lines for disposal, from 17% to 16%.

[5] Much of the waste destined for recycling may also have crossed state lines, but waste destined for recycling is not as controversial as that sent for disposal. In addition, recycling facilities do not generally require permits by state agencies. Thus, amounts shipped across state lines for recycling cannot generally be tracked by the solid waste agencies.

[6] See "Federal Appeals Court Strikes Majority of Virginia Restrictions on Trash Imports," *Daily Environment Report*, June 7, 2001, p. A-2. The case decided was Waste Management Holdings, Inc. v. Gilmore, 252 F.3d 316 (4th Cir 2001).

[7] Letter of Senators Stabenow and Levin to Hon. Laurel C. Broten, Ontario Minister of the Environment, August 30, 2006.

[8] Letter of Laurel C. Broten, Ontario Minister of the Environment, to Senators Stabenow and Levin, August 30, 2006.

[9] United Haulers Ass'n v. Oneida-Herkimer Solid Waste Mgmt. Auth., 127 S. Ct. 1786 (2007).

[10] Personal communication, New Jersey Department of Environmental Protection, May 30, 2007.

[11] Transfer stations receive waste from collection trucks, compact it, bale it, and load it on larger trucks for disposal elsewhere.

[12] Another Canadian province, British Columbia, also exports waste to the United States, but the amount is substantially smaller (about 100,000 tons to Washington state).

[13] Illinois, like most states, does not report waste exports. This export estimate was derived from data provided by neighboring states.

[14] "Waste Age 100," *Waste Age*, June 2006, p. 22.

[15] "The State of Garbage in America," *BioCycle*, April 1994, p. 51, and April 2006, pp. 38, 40.

In: Solid Waste: New Research
Editors: A. S. Martin and J. D. Suarez

SBN: 978-1-60456-809-7

Chapter 5

SOLIDIFIED/STABILIZED MUNICIPAL SOLID WASTE INCINERATION FLY ASHES IN CEMENT BASED MATRIX: ALUMINUM HYDROLYSIS EFFECT UPON LEAD SPECIATION IN SOLIDS

***Jacques Yvon*[1*]*, Diano Antenucci*[2]*, El-Aïd Jdid*[1]*, Gabriel Lorenzi*[2]*, Veronika Dutre*[3]*, Dany Leclerq*[2] *and Peter Nielsen*[3]**

[1] Nancy Université : Laboratoire Environnement et Minéralurgie (LEM), UMR 7569 CNRS - INPL, École Nationale Supérieure de Géologie, 15 Avenue du Charmois, B. P. 40, F-54 501 VANDŒUVRE-LES-NANCY - France

[2] Institut Scientifique de Service Public (ISSeP), Rue du Chéra 200, B-4000 LIEGE - Belgique

[3] Vlaamse Instelling voor Technologisch Onderzoek (VITO), Boeretang 200, B-2400 MOL - Belgique

ABSTRACT

Usually, the Municipal Solid Waste Incineration (MSWI) fly ashes are solidified/stabilized (S/S) in cement and disposed in appropriate landfills, producing huge monoliths. In order to understand the chemical reactions, which can take place inside such monoliths, an analytical approach at the micron and even at the infra-micron scale is required. Considering, this approach, which practically combines wet chemical analyses, leaching tests, XRD and SEM-EDS characterization, samples of naturally aged and fresh S/S MSWI fly ashes, were examined to describe and to predict the long-term behavior of lead. The compliance leaching tests indicate that considerable amounts of lead are leached for several samples. In addition to portlandite and calcite, the main major minerals identified are hydrocalumite, ettringite and ordered silicate-hydrates, the three later being free of lead in detectable concentrations. The lead speciation shows a rather large variety of minor Pb-bearing compounds, including soluble minerals and amorphous phases. As observed by SEM performed on polished thin sections, the neo-formed Pb-

* Correspondence to : (*) jacques.yvon@ensg.inpl-nancy.fr; Tel. +33 3 83 59 62 93; Fax +33 3 83 59 62 85

hydroxychlorides are partly related to a concentric sequence involving the aluminum hydrolysis, actually aluminum (internal) / aluminum hydroxide (intermediate) / hydrocalumite (external). Pb-minerals, typically cotunnite, blixite, laurionite, penfieldite or Pb-hydroxides are visible inside the hydrocalumite zone. Combined with the high Cl^- activity derived from the MSWI fly ashes, the aluminum hydrolysis occurring in cementing conditions partly governs the status of lead by promoting the precipitation of unstable neo-formed Pb-minerals.

Keywords: S/S MSWI fly ashes, cement, long-term Pb behavior, speciation, aluminum hydrolysis.

INTRODUCTION

In order to reduce the volume of municipal waste, high calorific wastes are incinerated with energy recovery. A variety of air pollution control (APC) technologies reduces the toxic materials emitted in combustion fumes. Among these, scrubbers neutralize acid combustion gases using a liquid spray, and dust filters remove tiny ashes from the gases. The ACP residues, i.e. the Municipal Solid Waste Incinerator (MSWI) fly ashes and residues generated by acid gas scrubbers, display rather high contents of chlorides, sulfates, and toxics as heavy metals, mainly Pb, Zn, Cr, Hg and Cd. Furthermore, leaching tests show that MSWI fly ashes are hazardous materials [1]. Considering that Western Europe produces about 1.2 10^6 tons/year of these materials [2], the MSWI fly ashes obviously generate an important eco-compatibility questions.

There are options for treating them prior to use or disposal. Treatment options include chemical extraction, thermal treatment (e.g. vitrification) and solidification/stabilization. According to the literature, the solidification/stabilization (S/S) technologies have been used for over three decades to treat various wastes [3]. Cement-based S/S process, in which waste materials are mixed with Portland cement and water, is often used to stabilize MSWI fly ashes [4]. Beyond the disposal of the solidified/stabilized MSWI fly ashes in appropriate landfills, producing huge “monoliths” which can encompass the entire disposal site called “monofills” [3], the question of potential utilizations of these solid products (fillers in concretes, material for ceramics,...) is more and more explored. As a consequence, the knowledge of the long-term behavior of solidified/stabilized MSWI fly ashes is a prerequisite. The essence of what we know at present about their evolution over time depends on data obtained at laboratory scale, by means of test-bars of small dimensions [5, 6] that quickly cool down. A difficulty still to be solved today consists in obtaining mobility models in “monofills”, where the porosities and the nature of phases evolve over time with the matter transfers and the reaction series strongly affected by a prolonged stage at a temperature which can reach 80°C [1], due to an adiabatic behavior of the huge monoliths. These reactions are of major significance since they govern the behavior and status of several pollutants, namely the inorganic elements such as heavy metals.

The understanding of the chemical reactions, which can take place inside a monofill, requires a petrologic approach at the micron and even at the infra-micron scale. Applied to aged monofills and fresh monoliths, such an approach leads to a general speciation of lead and shows the role of local reactions occurring during the chemical alteration of the S/S materials, typically the aluminum hydrolysis. The present paper is focused on the question of lead, one of the most dangerous and abundant metal contained in MSWI fly ashes. Perspectives on S/S MSWI fly ashes are also examined.

MATERIALS AND METHODS

Sampling

The study is based on the sampling of two landfill sites. One was 10 years old (A_1), where the MSWI fly ashes were stabilized with Portland cement, without any previous treatment, and the other was 5 years old (B_1), where the MSWI fly ashes were stabilized with cement after a partial leaching of water-soluble species, such as chlorides, and sulfidation of heavy metals using Na_2S.

Additionally to the samples taken from the two landfill sites, the filter fly ashes (A_{0f}) and the gas scrubber salts (A_{0s}) were collected separately in the incineration plant corresponding to the site A1, and the raw residues (B_0), an admixture of fly ashes and salts, and the fresh solidified MSWI residues (B_2) were sampled in the incineration plant corresponding to the site B_1.

The detailed methodology of sampling was reported previously [7].

Chemical Analyses

The total concentration of lead was determined using ICP-AES, after microwave digestion in an acid mixture of HNO_3, HBF_4, H_3PO_4, HCl and HF.

Availability Test

The NEN 7341 availability test is the Dutch standard extraction test for the assessment of maximum leachability [8]. The material was size reduced in a jaw crusher (< 10 mm) followed by grinding in a RETSCH centrifugal ball mill (S1000) until the material passed a sieve of 125 µm. The material (<125 µm) was then leached in two steps. In the first step it was leached at a liquid to solid ratio (L/S) of 50 L/kg with de-ionized water under pH static conditions (pH = 7, kept constant by feed-back control and addition of HNO_3 or NaOH) during 3 hours. In the second step, after filtration, the solid was leached at L/S = 50 at a pH of 4 (under pH static conditions) during another 3 hours. The two leachates were combined prior to analysis.

Compliance Test

The EN 12457 is an European standard test developed for compliance testing to determine whether a waste complies with specific reference values [9]. This test is one of the leaching tests described in annex II of the landfill directive 1999/31/EC (criteria and procedures for the acceptance of waste at landfills). The EN 12457 compliance test provides information on leaching of granular wastes and sludges at a liquid to solid ratio of 10 L/kg dry matter. The leaching values can be compared to European leaching limit values for acceptance of waste in landfills. These limit values would have been effective in the EU member states on the 1st of July 2005.

The four procedures described in this standardized test are based on different liquid to solid ratios and different particle sizes. Procedure 4 of the test was chosen because the required size reduction is the less severe (i.e. the particle size of 95% of the mass has to be < 10 mm compared to < 4 mm for the other 3 procedures). The particle size reduction is limited in order to maintain the physical state of the waste, thus the test conditions in procedure 4 more closely reflect the field situation. The leaching liquid is de-ionized water, and the contact time is 24 hours, during which the bottle containing the solid sample and leaching liquid is placed in an end-over-end-tumbler. The pH of the leachate is imposed by the leached waste. That of the raw fly ashes was between 12.4 and 12.5, and that of the solidified fly ashes was between 11.6 and 12.8.

Ph Static Tests

In this study a procedure proposed by CEN/TC 292 WG 6 (Nov. 1999) – basic characterization was used. The test was carried out with a fixed liquid to solid ratio of 10 on material with a particle size <1 mm, over 50 hours test duration. Starting with a contact period of 2 hours, in which steady state conditions were reached for most constituents, followed by a pH control period of 44 hours and ending with a steady state verification period of 4 hours. The tests were performed at seven fixed pH values between 4 and 12, and at natural pH without any adjustment. They were carried out in a titration device, where the pH was continuously measured and adjusted with either nitric acid or sodium hydroxide. The metal concentration in the elution solutions was analyzed through ICP-AES.

X-Ray Diffraction Analyses

Identification of major mineral phases were carried out by X-ray diffraction using a Philips PW 1390 diffractometer equipped with monochromatized Fe$K\alpha$ radiation. Attempt at identifying minor mineral phases by X-ray diffraction analyses using a Gandolfi camera has failed. Practically, no reliable results were obtained for minerals with crystal size below 80 μm. This situation is particularly problematic with the Pb-bearing compounds that are about 1 μm in size.

Scanning Electron Microscopy (SEM)

Due to the very small size of Pb-bearing phases present in the solidified MSWI fly ashes, the identification of the various Pb-compounds was realized using a Hitachi S-2500 SEM, equipped with a "superquantum DELTA" EDS micro-analyser. The preliminary identification was performed using a "K.E. Development" Back Scattered Electron Imaging device.

The SEM analyses of solidified MSWI were performed on carbon vaporized, polished thin section of 30 μm. Though the rock thin sections technique is well known, the specificity of the materials analysed here needs adapted care. The intrinsic characteristics of solidified MSWI required a specifically adapted preparation due to their low cohesion, the potential toxicity of their sawn products and the high solubility or the limited thermal stability of some components.

RESULTS

Chemical Analyses and Leaching Tests

For each group of samples (A_{0f}, A_{0s}, A_1, B_0, B_1 and B_2), the lowest and the highest total and leached lead concentrations are summarized in table 1. The limit values defined by European directive [10] are given as well. The results of pH static tests are given in table 2.

Table 1. Total Pb concentration (mg/Kg) and leached amounts (mg/Kg) from raw and solidified MSWI fly ashes. Symbols are as follow: A_{0f} = raw MSWI fly ashes; A_{0s} = raw gas scrubber salts; A_1 = 10 years old solidified residues A_{0f} and A_{0s}; B_0 = raw MSWI fly ashes and salts ; B_1 = 5 years old solidified residues B_0; B_2 = fresh solidified residues B_0; Eu = European regulation concerning Pb compliance test values admitted at L/S = 10 L/Kg for inert wastes*, not dangerous wastes and dangerous wastes*** [10].**

	Total concentration (mg/kg)	Availability tests (mg/kg)	Compliance tests (mg/kg)
A_{0f}	5150 - 7010	640 - 780	0.13 – 0.74
A_{0s}	215 - 236	27 - 32	18 - 19
A_1	425 – 2750	185 – 1032	1.28 – 9.24
B_0	3710	611	820
B_1	1060 – 1690	51 – 257	$\leq$0.11
B_2	1430 – 1500	133 – 207	5.9 – 7.7
Eu			0.5* 10** 50***

As indicated by the results, the total lead content is rather high in both raw MSWI residues and their solidification/stabilization products.

The relative availability, defined as the ratio of maximum availability to the total concentration, shows values slightly lower for filter fly ashes A_{0f} ($\cong$ 11%) and for gas scrubber salts A_{0s} ($\cong$ 13%) than for the raw MSWI residues B_0 ($\cong$ 16%). The solidification/stabilization products of A_{0f} and A_{0s}, actually the A_1 samples, show the highest relative availability (≅ 40%), whereas the lowest values are obtained from the S/S products of B_0, actually the B_1 and B_2 samples (≅ 5 to 15%).

Table 2. Results of pH static tests, given in mg/kg. Same symbols as for Table 1.

pH	A_{0f}	A_{0s}	A_1	B_1	B_2
4	23.24	4.84	0.19-0.47	< 0.10	< 0.10
5.5	0.77	2.21	0.13-0.37	< 0.10	< 0.10
7	0.15	1.09	0.11-0.35	< 0.10	< 0.10
9	0.12	2.66	0.12-0.43	< 0.10	< 0.10
10	0.20	2.49	0.11-0.30	< 0.10	< 0.10
11	0.25	5.33	0.13-8.47	< 0.10	< 0.10
12	79.81	9.43	0.43-52.46	< 0.10	0.34

Considering the compliance test data, it emerges that more than 60% of the available lead are leached from the A_{0s} gas scrubber salts, and that the totality of the available lead is removed from the B_0 MSWI residues. The lowest leached amount is obtained with the A_{0f} filter fly ashes whose leachate also displays the lowest pH. The highest leached concentrations are those measured from the S/S products labelled A_1 and B_2 that also reveal the highest leachate pH. When compared with the European landfill directive, the compliance test results show that most of the solidified MSWI fly ashes cannot be considered as inert wastes but they should be considered as dangerous wastes.

The pH static tests (pH = 4 to 12) on the raw and solidified MSWI residues show that the leaching of lead is strongly pH dependent. The leachates with the lowest concentrations occur between pH = 6 and 10. The highest leached amounts of lead are found at pH ≥12.

X-Ray Diffraction Analyses

The data (table 3) indicate the typical mineralogy of MSWI residues. In addition to calcite and, in some cases, portlandite, the S/S products systematically contain hydrocalumite and ettringite. In case of aged and fresh solidified samples (respectively A_1 and B_2) katoite and tobermorite are also present.

Table 3. XRD analyses of raw and S/S MSWI residues. Same symbols as for tables 1 and 2

A_{0f}	A_{0s}	A_1	B_0	B_1	B_2
Calcite Anhydrite Quartz Sylvite Halite Pyrite hematite	Calcite Portlandite $CaCl_2Ca(OH)_2$- H_2O (carbone)	Calcite Portlandite Hydrocalumite Ettringite katoite	Calcite Portlandite Quartz $CaCl_2Ca(OH)_2$- H_2O Anhydrite Fluorite Halite sylvite	Calcite Portlandite Hydrocalumite Ettringite	Calcite Hydro- calumite Ettringite Katoite tobermorite

Sem Analyses

Lead Speciation.

The SEM images and EDS analyses show that katoite, ettringite and calcium silicate hydrates, namely major compounds that can potentially incorporate heavy metals, do not contain any detectable lead. The rather large variety of Pb-minerals and Pb-bearing minerals and amorphous phases identified in raw and solidified MSWI fly ashes is reported in table 4.

Table 4. Lead speciation in raw and solidified MSWI residues. Symbols are the same as in previous tables.

Family	Mineral species - like compositions	Chemical formula
	A_{0f} and A_{0s} samples	
Chlorides	Cotunnite	$PbCl_2$
Oxychlorides	Blixite	$Pb_2Cl_{2-x}(O,OH)_{2x}$
	Penfieldite	$Pb_2Cl_3(OH)$
	Laurionite	PbClOH
Phosphates	Hydoxyapatite	$(Ca_{1-x}Pb_x)_{10}(PO_4)_6(OH)_2$
Vanadates	Chervetite	$Pb_2V_2O_7$
	A_1 samples	
Chlorides	Cotunnite	$PbCl_2$
Oxychlorides	Blixite	$Pb_2Cl_{2-x}(O,OH)_{2x}$
	Penfieldite	$Pb_2Cl_3(OH)$
Chlorides-hydroxides	-	$Pb_2CuCl_2(OH)_4$
Hydroxides	-	$Pb(OH)_2$
Carbonates	Hydrocerusite	$Pb_3(OH)_2(CO_3)_2$
Metal	Lead	Pb
Disordered Calcium Silicate	CSH	Pb-CSH
Hydrates	-	-
Glassy phases		
	B_0 Samples	
Chlorides	Cotunnite	$PbCl_2$
Glassy phases	-	-
Calcium aluminum silicates	Complex phases	Pb-(Ca-K-Al-Si)
	B_1 samples	
Chlorides	Cotunnite	$PbCl_2$
Oxides	-	PbO
Hydroxides	-	$Pb(OH)_2$
Sulfides	Galena	PbS
Sulfates hydroxides	-	$Pb_4Cu(SO_4)_2(OH)_6$
	B_2 samples	
Chlorides	Cotunnite	$PbCl_2$
Oxides	-	PbO
Hydroxides	-	$Pb(OH)_2$
Oxides hydroxides	-	$Ca_2Pb_2O_5(OH)_2$
Sulfides	Galena	PbS
Metal	Lead	Pb

Aluminum Hydrolysis

SEM images and EDS spectra obtained from solidified MSWI fly ashes are illustrated in figure 1. Whatever the age of the solidified MSWI fly ashes, most of the SEM images and EDS analyses systematically show a concentric sequence: aluminum (internal) / aluminum hydroxide (intermediate) / hydrocalumite (external) (figure 1).

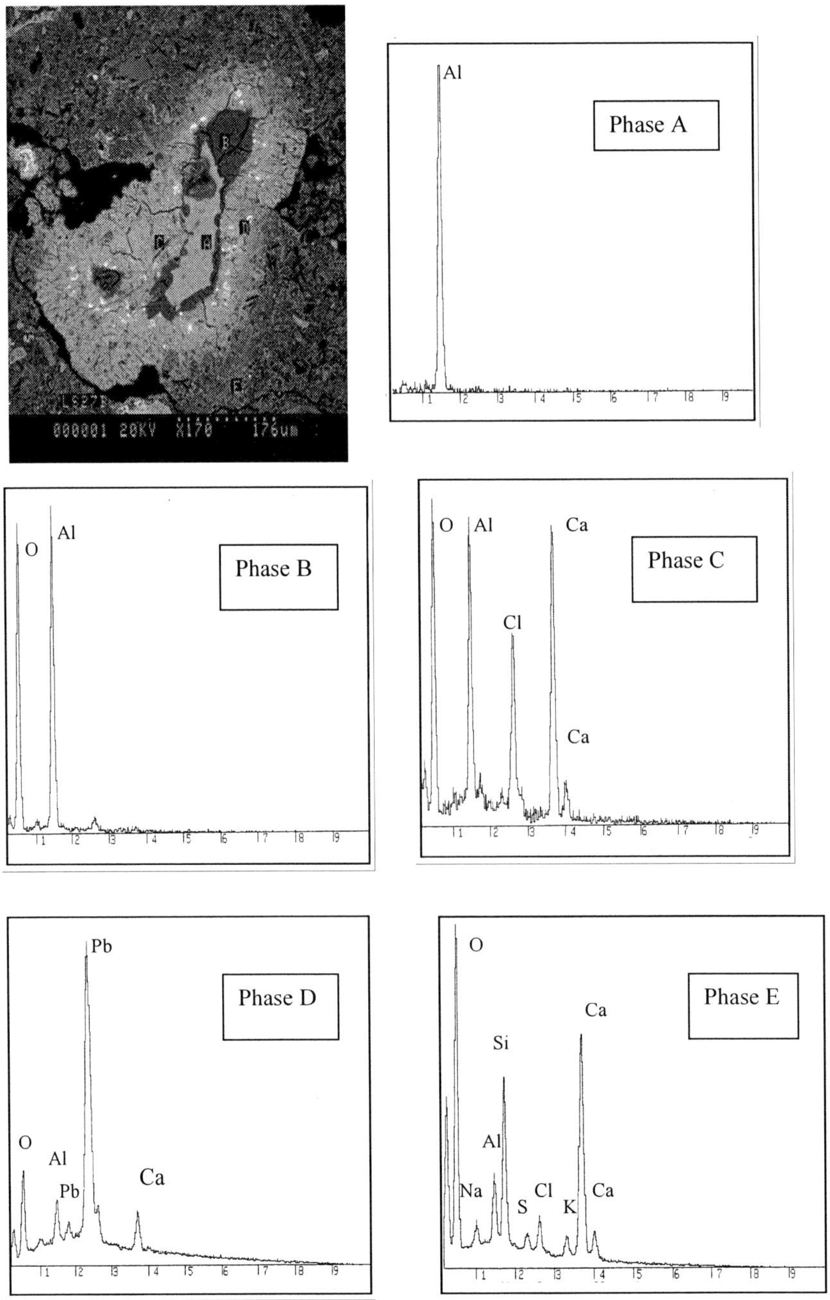

Figure 1. SEM images and corresponding EDS spectra of aluminium hydrolysis by-products. A = aluminium; B =gibbsite, $Al(OH)_3$, and bœhmite or diaspore, $AlO(OH)$; C = hydrocalumite, $Ca_2Al(OH)_6(Cl_{1-x}(OH)_x)\cdot 3H_2O$; D = blixite (white particles), $Pb_2Cl_{2-x}(O, OH)_{2x}$.

Depending on the position of the section, hydrocalumite alone, hydrocalumite and aluminum hydroxide or the three phases together are observed. Spots of Pb-minerals, typically cotunnite, blixite, laurionite, penfieldite or Pb-hydroxides are visible inside the hydrocalumite zone (figures 1 and 2). It must be stressed that the cationic content of hydrocalumite is limited to Ca^{2+} and Al^{3+}. In samples characterized by a chloride depletion, total exchange of Cl^- for SO_4^{2-} anions is observed, thus corroborating the well known structural anion exchange capacity resulting from the excess of cationic charges in the framework of layered double hydroxides (LDHs) such as hydrocalumite [11, 12].

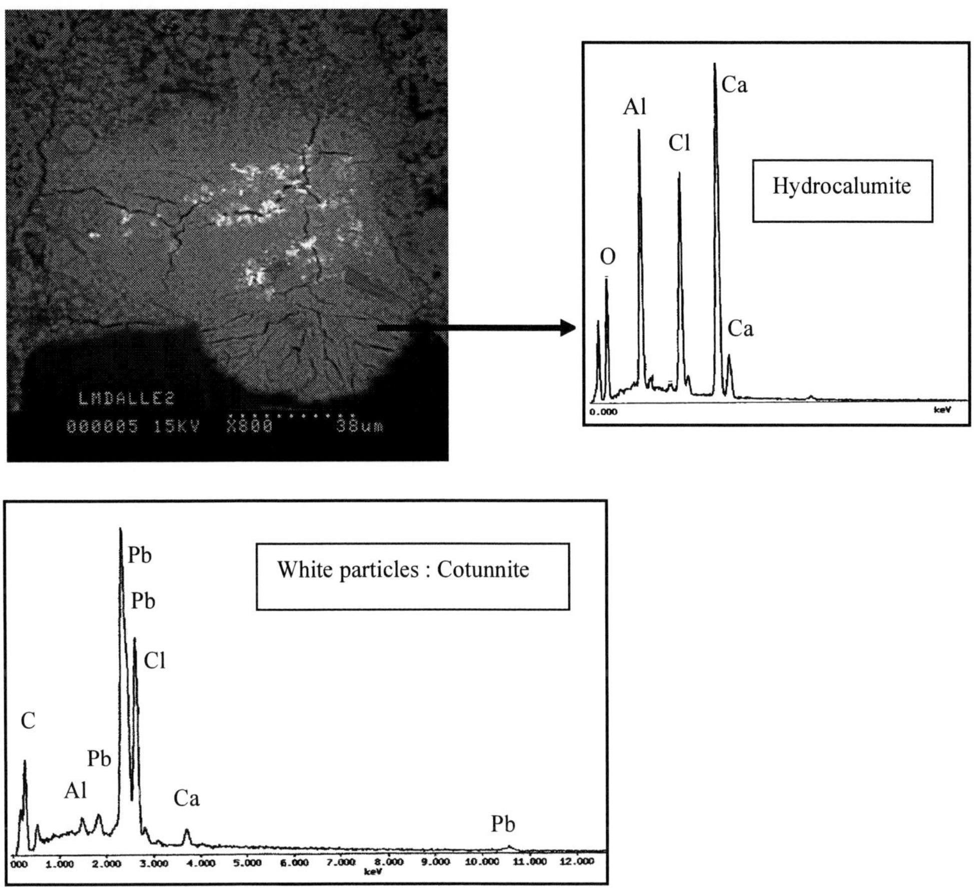

Figure 2. SEM images and corresponding EDS spectra of hydrocalumite and cotunnite (white particles)

Aluminum hydrolysis also occurs in the filter fly ashes A_{0f}, and several Pb-minerals already mentioned above are observed in the hydrolysis zone. However, the hydrocalumite is missing there.

DISCUSSION

Chemical Analyses and Leaching Tests

The whole leaching results (tables 1 and 2) show that both raw and solidified fly ashes display a strong pH dependence and are more prone to leaching in strongly alkaline conditions than in the neutral conditions or those of the maximum availability test (pH = 4). However, it is difficult to compare the results of both the tests, due to many factors, for instance : the size distribution of the particles, which is specific to each test (below 125 μm for the availability test, 10 mm for the compliance test and 1 mm for pH static test), and the heterogeneity of the samples that also depends on grain sizes. Nevertheless, due to its amphoteric behavior, lead may be more prone to leaching in the compliance test than in the availability tests, especially when the material imposes a very high pH (> 12) on the leachates.

The fact that the solidified MSWI fly ashes reveal leached concentrations of lead lower than the raw MSWI residues cannot be explained by the mineralogy of lead, similar in both the cases. Taking into account the above description of the Al hydrolysis occurring in S/S MSWI residues, it can be stated that the lowering of lead leached level could mainly be due to a temporary screen effect of the hydrocalumite zone.

X-Ray Diffraction Analyses

Except for calcite that derives both from neutralization reagents and a carbonation product of portlandite, no major mineral phases typical of the MSWI residues were found in their solidified products. The neo-formed major compounds (hydrocalumite, katoite, tobermorite) give several indications on huge monoliths of S/S MSWI residues. In nature, Si-rich hydrothermal solutions are known to promote the crystallization of silicate-hydrates such as katoite and torbermorite associated with hydrocalumite [13]. Recent experimental studies show that layered double hydroxides (LDHs), characterized by a structure-type referring to hydrocalumite, have been synthesized in hydrothermal conditions at temperatures ranging between 65 and 110°C [14]. Furthermore, it must be stressed that the temperature measured was close to 70°C at one-meter depth when sampling the five years old landfill (samples B_1). Thus, in huge monolith of S/S MSWI residues, the crystallization of ordered silicate-hydrates (katoite, torbermorite) and hydrated calcium aluminates (hydrocalumite), is promoted by hydrothermal fluid circulation and temperatures exceeding 70°C.

SEM Analyses

Interpretation of Aluminum Hydrolysis by-Products Sequence

Considering the textural image observed under SEM analyses, the various by-products of the aluminum corrosion occurring in the solidified MSWI residues are interpreted as follow: the hydrolysis of aluminum produces aluminum hydroxides, inducing a local pH decrease according to the global reactions (I) and (II) respectively producing gibbsite or bayerite and bœhmite or diaspore:

$$Al + 3(OH)^- \rightarrow Al(OH)_3 + 3e^- \quad \text{(I)}$$

$$Al + 2(OH)^- \rightarrow AlO(OH) + H^+ + 3e^- \quad \text{(II)}$$

The excess of aluminum still present in the solution under the form of aluminate precipitates as hydrocalumite, according to the schematic reaction (III), which occurs at lower pH since it needs protons,

$$2Ca(OH)_2 + Al(OH)_4^- + Cl^- + 2H^+ + H_2O \rightarrow Ca_2Al(OH)_6Cl\text{-}3H_2O \quad \text{(III)}$$

The required ions: calcium, chloride (and/or sulfate) being available from Portland cement (calcium and sulfate source) and MSWI fly ashes (calcium, chloride and sulfate source). The lead minerals connected to this sequence precipitate after the beginning and before the end of hydrocalumite formation; they also need protons to precipitate according to reactions (IV) to (VII), their solubility being quickly reached when the pH decreases:

$$Pb(OH)_3^- + H^+ \rightarrow Pb(OH)_2 + H_2O \quad \text{(IV)}$$

$$Pb(OH)_3^- + 2H^+ + Cl^- \rightarrow PbClOH + 2H_2O \quad \text{(V)}$$

$$Pb(OH)_3^- + 3H^+ + 2Cl^- \rightarrow PbCl_2 + 3H_2O \quad \text{(VI)}$$

$$2Pb(OH)_3^- + 5H^+ + 3Cl^- \rightarrow Pb_2Cl_3(OH) + 5H_2O \quad \text{(VII)}$$

The origin of lead compounds, which precipitate into the hydrocalumite zone, is rather intriguing. It could be stated that, initially, the lead occurs from condensation of vapours (e.g. metal or chlorides) since lead condensations can be detected around fly ashes particles (figure 3) and that secondarily, lead substitutes to calcium in calcium compounds during stabilization process and finally is expelled under the form of plombate anions from these structures when the re-crystallization process occur, due to hydrothermal evolution. Nevertheless, these condensation figures are rarely observed and the aluminum external surface does not show more lead contamination than other particles. Published data indicate the presence of lead traces (up to 70 ppm) in various aluminum alloys [15], but this amount is not sufficient to generate the total occurrence of lead compounds identified. Then, it can be proposed that the anionic forms of lead (e.g. $Pb(OH)_3^-$) are present in the S/S material alkaline interstitial solutions prior to the aluminum corrosion and precipitate into diverse solid compounds (eq. IV to VII) in a range of pH close to the precipitation field of hydrocalumite.

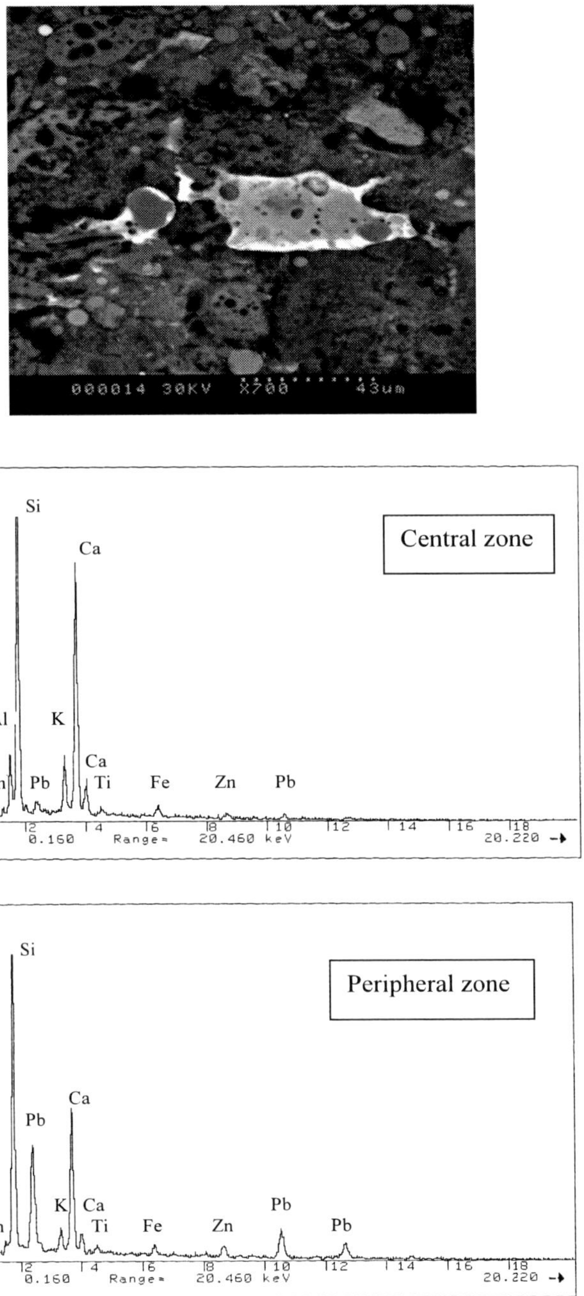

Figure 3. Pb condensation around fly ashes particles; the peripheral zone is enriched in Zn and Pb.

In the case of Al hydrolysis occurring in the MSWI residues, in particular in the filter fly ashes A_{0f}, the sequence of crystallization is simplified. Indeed, the step (III) described above and leading to the precipitation of hydrocalumite is missing. Considering the discussion about the major neo-formed minerals, the absence of hydrocalumite-like phase may be the consequence of an Al hydrolysis process that takes place at room temperature, actually too low for the crystallization of hydrated calcium aluminates compounds with LDHs-type structure.

Status of Lead

Basically, the solidification/stabilization of MSWI fly ashes using cement causes the precipitation of several Pb-minerals, such as oxides, hydroxides, chlorides, hydroxychlorides, oxychlorides and, in some cases, sulfides that are logically observed after hydrometallurgical treatment of MSWI using Na_2S (Table 4). Some lead chlorides and oxychlorides are inherited from the raw MSWI fly ashes, but some others are neo-formed during and/or after the S/S process. This is actually the case for the Pb-chlorides, Pb-hydroxychlorides and Pb-oxychlorides related to the aluminum hydrolysis by-products sequence. Except for CSH and some calcium aluminum silicates, no example of lead involving cationic substitutions was observed. In particular, the EDS spectra do not show any detectable lead in ettringite, katoite or hydrocalumite. Various divalent cations can substitute for Ca^{2+} in ettringite, but Pb^{2+} is not mentioned [5]. In garnets-type structure, Pb^{2+} can replace Ca^{2+} only in the vanadate-substituted garnets [16]. Despite the fact that various divalent cations can fit into the structure of LDHs (17, 18), the presence of lead in these compounds has never been reported. Usually, the ionic radii are supposed to play a critical role in the cationic substitution mechanisms. Ca^{2+} is 7-fold coordinated in hydrocalumite [11] and 8-fold coordinated in ettringite [19] and katoite [20]. The ionic radii reported by Shannon [21] for Ca^{2+} and Pb^{2+} in 7- and 8-fold coordination are significantly different: $^{[VII]}Ca^{2+}$ = 1.06; $^{[VII]}Pb^{2+}$ = 1.23; $^{[VIII]}Ca^{2+}$ = 1.12; $^{[VIII]}Pb^{2+}$ = 1.29. Nevertheless, the situation does not allow any exhaustive explanation for the lack of Pb^{2+} substitution for Ca^{2+} observed in ettringite, katoite and hydrocalumite. Indeed, Ca-minerals that accept Pb^{2+} substitution for Ca^{2+} in their structure exist, for example apatite [22, 23] and tsumcorite [24] groups. In addition to the ionic size criterion, the stereochemical active $6s^2$ lone-pair behavior for Pb^{2+} must be considered as well. In general, the effect of ns^2 lone-pair (n = 5, 6) causes remarkable polyhedral distortions, which are maximized in case of chemical bonds with high ionic character [25]. In the case of Pb^{2+}, the situation is observed in various Pb-hydroxychlorides, for instance those displaying the common thorikosite-type structure, $Pb_3SbO_3(OH)Cl_2$, which includes, among others, blixite [26]. In these minerals, the active $6s^2$ lone-pair effect often manifests itself by decreasing the average Pb-O bond lengths and increasing the average Pb-Cl bond lengths in the Pb-sites [27]. The distortion of a given crystallographic site can be estimated by mean of bond-length distortion (BLD). According to Wenger and Armbuster [28], the bond-length distortion (BLD) is defined as follows:

$$BLD(\%) = \frac{100}{n}\sum_{i=1}^{n}\frac{|d_i - d_m|}{d_m}$$

where n = number of cation-anion bonds, d_i = individual cation-anion bond lengths and d_m = mean cation-anion bond-length. Ideally, the edge-length distortion (ELD) would be also considered but the individual edge lengths are often missing from published crystallographic data. BLD was calculated for Ca- and Pb-sites of minerals identified in the S/S MSWI fly ashes, namely ettringite [19], katoite [20], hydrocalumite [29], cotunnite [30], penfieldite [31] and pinalite [26] (bond distances for blixite was not found).

Considering the coordination number for Ca^{2+} in hydrocalumite, katoite and ettringite, BLD have been also calculated for some Ca-minerals where Pb^{2+} substitutes for Ca^{2+} in polyhedra involving coordination numbers 7 or 8, namely fluorapatite, $Ca_5(PO_4)_3F$ [32], hydroxylapatite, $Ca_5(PO_4)_3OH$ [32], cabalzarite, $Ca(Mg,Al,Fe)_2(AsO_4)_2(H_2O,OH)$ [24] and tsumcorite *s.s.*, [33]. The compared BLD values reported in table 5 are rather relevant. Among the minerals identified in the S/S MSWI fly ashes, only those displaying polyhedra with high BLD values accept Pb^{2+} in their structure, actually the Pb-chlorides and Pb-oxychlorides. Furthermore, the selected Ca-minerals where Ca^{2+} can be replaced by Pb^{2+}, reveal (Ca^{2+}, Pb^{2+})-polyhedra with BLD values drastically higher than those obtained for Ca-polyhedra in ettringite, katoite and hydrocalumite. The results support the idea that, despite the ionic radii difference, Pb^{2+} substitution for Ca^{2+} is allowed in Ca-minerals that can accommodate highly irregular polyhedra. As a consequence, one can assume that in addition to the ionic size, the effect of the $6s^2$ lone-pair combined with the high activity of Cl^- and the high pH, typical for S/S MSWI fly ashes materials, promote the precipitation of Pb-minerals characterized by atomic structures that tolerate distorted crystallographic sites. Thus, the physicochemical conditions combined with the crystallochemical constraints are responsible for the crystallization of Pb-bearing minerals that may not ensure the immobility of lead at high pH as already shown by the pH static leaching tests (see above).

Table 5. Bond-length distortion (BLD) values for Ca and Pb-sites in different minerals

Mineral	Coordination polyhedron	BLD (%)
Ettringite	$^{[VIII]}Ca(1)$	1.7
	$^{[VIII]}Ca(2)$	2.2
Katoite	$^{[VIII]}Ca(1)$	0
	$^{[VIII]}Ca(2)$	0
Hydrocalumite	$^{[VII]}Ca$	1.6
Cotunnite	$^{[IX]}Pb$	7.1
Penfieldite	$^{[V]}Pb(1)$	7
	$^{[V]}Pb(2)$	8.8
Pinalite	$^{[VIII]}Pb$	13
	$^{[IX]}Pb$	4.4
Fluorapatite	$^{[VII]}Ca$	4.5
	$^{[VII]}Pb$	4.7
Hydroxylapatite	$^{[VIII]}Ca$	3.9
	$^{[VIII]}Pb$	7.2
Cabalzarite	$^{[VIII]}Ca$	7.9
Tsumcorite *s.s.*	$^{[VIII]}Pb$	6.9

CONCLUSION

Several Relevant Points May Be Stressed from the Study

In huge monolith of S/S MSWI residues, hydrothermal fluids circulation at $T \geq 70°C$ promotes the crystallization of ordered silicate-hydrates (hydrogarnets) and hydrated calcium aluminates (hydrocalumite).

The hydrocalumite formed in these hydrothermal conditions is connected to the Al (or other amphoteric metals) hydrolysis. As a consequence, the sequence of Al hydrolysis observed in S/S MSWI residues reveals an additional step with regard to the hydrolysis sequence occurring in the raw MSWI residues where hydrocalumite is missing.

The combination of sterical effect and physicochemical conditions contribute to the precipitation of Pb-chlorides and Pb-hydroxychlorides, lead being expelled from the initial calcium containing minerals, thus explaining why practically no cationic substitution involving lead was observed in the various neo-formed ordered hydrated compounds.

Considering that the Pb-minerals are embedded in the hydrocalumite zone, the leached amounts of lead from S/S MSWI residues is only temporary reduced with regards to the raw MSWI residues. Taking into account the high solubility of Pb-chlorides and Pb-oxychlorides at $pH \geq 12$, the removal of lead will depend on the accessibility and solubility of hydrocalumite at high pH conditions. A situation that may not ensure the long-term stability of lead.

Outlook

The present state of things does not allow any re-use of MSWI residues solidified/stabilized in cement matrix. In addition to the mobility of lead discussed here, other published data on these solid materials have also revealed drastic leached concentrations of Cl^- as well as rather high mobility of SO_4^{2-} (7). As a consequence, S/S MSWI residues must be disposed of in landfills for dangerous wastes.

In order to cope with the present situation, several improvements may be suggested for reducing the toxicity of S/S MSWI residues with regard to Pb, Cl^- and SO_4^{2-}.

Prior to the solidification/stabilization of MSWI in cement, lead would be better immobilized in phosphate mineral phases displaying a rather high stability such as apatite-like or crandallite-like structure compounds.

By analogy, the question of the drastic mobility of Cl^- could be solved by the precipitation of stable chlorapatite.

The precipitation of $BaSO_4$, for example from precipitated barium carbonates as usual in roof tile technology, could be a solution for a better stabilization of SO_4^{2-}.

The last point is that for avoiding temperature increases after the deposition, the landfill must be designed either in small alveoli or thin layers.

ACKNOWLEDGEMENTS

Authors are grateful to LIFE for financial support (Program Research LIFE99 ENV/B/000638).

REFERENCES

[1] Antenucci D., Barrès O., Dreesen R., Dutré V., Jdid E.A., Kenis C., Leclerq D., Lhote F., Lorenzi G., Nielsen P., Veschkens M., Yvon J. (2003). Evaluation du comportement à long terme des résidus fins d'incinération des ordures ménagères (REFIOM) stabilisés aux liants hydrauliques ; LIFE99 ENV/B/000638, Final report, 35pp.

[2] ASSURE. (2000). Incineration in Europe; *http://www.assure.org/currentissues /incinerationprofile.ppt*

[3] Barth E.F., de Percin P., Arozarena M.M., Zieleniewski J.L., Dosani M., Maxey H.R., Hokanson S.A., Pryately C.A., Whipple T., Kravitz R., Cullinane M.J., Jones L.W., Malone P.G. (1990). Stabilization and Solidification of Hazardous Wastes; *Pollution Technology Review*, No 186, ndc, New Jersey. 390pp.

[4] Robert H. (1995). Congrès International sur les procédés de solidification et de stabilisation des déchets ; *Nancy ; Proceedings*. 223-227.

[5] Lin T.T., Lin C.F., Wel W.C., Lo S.L. (1993). *Environ. Sci. Technol.* 27, 1312-1318.

[6] Brault S. (2001). Modélisation du comportement à la lixiviation à long terme de déchets stabilisés à l'aide de liants hydrauliques ; PhD ; Université Pierre et Marie Curie, Paris VI ; 266pp.

[7] Yvon J., Antenucci D., Jdid E.A., Lorenzi G., Dutré V., Leclerq D., Nielsen P. & Veschkens M. (2006). *J. of Geochemical Exploration*, 90, 143-155.

[8] NEN 7341 (1995). Leaching characteristics of solid earthy and stony building and wastes materials. Leaching tests. Determination of the leaching of inorganic components. 19p.

[9] PREN 12457/4 (1999). Characterization of waste. Leaching. Compliance test for leaching of granular waste materials and sludges – Part 4: one stage batch test at a liquid to solid ratio of 10 l/Kg with particle size below 10 mm (without or with size reduction). 19pp.

[10] Projet de décision de la Commission établissant des critères et des procédures d'admission de déchets dans les décharges, conformément à l'article 16 et à l'annexe II de la directive 1999/31/CE concernant la mise en décharge des déchets. 2002, 31p.

[11] Kalinichev A.G., Kirkpatrick R.J., Cygan R.T. (2000). *Amer. Min.* 85, 1046-1052.

[12] Leroux F.; Besse J-P. (2001). *Chem. Mater.* 13, 3507-3515.

[13] Passaglia E., Rinaldi R. (1984). *Bull. Min.* 107, 605-618.

[14] Möhmel S., Kurzawski I., Uecker D., Müller D., Gessner W. (2002). *Cryst. Res. Technol.* 37, (4), 359-369.

[15] Maystre L.Y., Duflon V. (1994). Déchets urbains : Nature et caractérisation. Presses Polytechniques et Universitaires Romandes. Collection Gérer l'Environnement. p. 175.

[16] Khorari S. (1997). Cristallochimie des arséniates de structure alluaudite. PhD. Université de Liège. 262p.

[17] Scheidegger A.M., Wieland E., Scheinost A.C., Dähn R., Spieler P. (2000). *Environ. Sci. Techol.* 34, 4545-4548.

[18] O'Hare D. & Khan Aamir I. (2002). *J. Mater. Chem.* 12, 3191-3198.

[19] Moore A.E., Taylor H.F.W. (1970). *Acta Cryst.* B26, 386-393.

[20] Sacerdoti M. & Passaglia E. (1985). - *Bull. Min.* 108, 1-8.

[21] Shannon R.D. (1976). Revisited effective ionic radii and systematic studies of interatomic distances in halides and chalcogenides. *Acta Cryst.* A32, 751-767.

[22] Shevade A.V., Erickson L., Pierzynski G., Jiang S.J. (2001). *Hazardous Subst. Res.* 3, 2-12.

[23] Leroy N. & Bres E. (2001). *European Cells and Materials*, 2, 36-48.

[24] Brugger J., Meisser N., Schenk K., Berlepsch P., Bonin M., Armbruster T., Nyfeler D., Schmidt S.(2000). *Amer. Min.* 85, 1307-1314.

[25] Kyono A., Kimata M., & Shimizu M. (2000). *Amer. Min.* 85, 1287-1293.

[26] Grice J.D. & Dunn P.J. (2000). *Amer. Min.* 85, 806-809.

[27] Welch M.D., Cooper M.A., Hawthorne F.C., Criddle A.J. (2000). *Amer. Min.* 85, 1526-1533.

[28] Wenger M., Armbruster T. (1991). *Eur. J. Mineral.* 3, 387-399.

[29] Sacerdoti M., Passaglia E.N. (1988). Jb. Miner. Mh. H. 10, 462-475.

[30] Chichagov A.V., Belonozhko A.B., Lopatin A.L., Dokina T.N., Samokhvalova O.L., Ushakovskaya T.V., Shilova, Z.V. (1990). *Kristallographiya*, 35, 610-616.

[31] Merlino S., Pasero M., Perchiazzi N., Gianfagna A. (1995). *Mineral. Mag.* 59, 341-347.

[32] Hughes J.M., Cameron M. and Crowley K.D. (1989). *Amer. Mineral.*, 74, 870-876.

[33] Krause W., Belendorff K., Bernhardt H., McCammon C., Effenberger H. and Mikenda, W. (1998). *Eu. J. Mineral.*,10, 179-206.

In: Solid Waste: New Research
Editors: A. S. Martin and J. D. Suarez

SBN: 978-1-60456-809-7

Chapter 6

Spatial Modeling and Optimization of Municipal Solid Waste Collection in Urban Regions

Lubos Matejicek
Institute for Environmental Studies
Charles University in Prague, Faculty of Natural Science
Benatska 2, 128 01, Prague, Czech Republic

Abstract

Growing production of municipal solid waste in the recent years caused by increasing global levels of economic activity needs more sophisticated management tools for its collection. Despite some local efforts to reduce the waste, disposal in landfills is still the most predominate way. Considering to restrictive environmental regulations, more efficient and sustainable systems of municipal solid waste management are carried out to protect our living environment. They deal with the waste sorting, which is mainly focused on categories represented by paper, mixed plastics, mixed glass, mixed waste, hazardous waste and metals. Thus, specific transport vehicles and optimized routes are needed to manage the collection of all the waste categories from local sites to landfills, incineration plants, collecting yards and recovery sites. This contribution presents the waste collection methods based on spatial analysis and operations research. The spatial analysis is focused on development of the road network datasets for waste collection, finding the best routes, defining the service area and creating the origin-destination cost matrix in geographic information systems (GISs). The identified routes and local service areas together with the cost per unit and the total amount of the transported waste form data inputs into operations research. The transportation problem in linear programming is used to optimize the total cost for transportation of each waste category in the framework of municipal solid waste collection. As an example, the urban areas of Prague are included into a case study to illustrate spatial modeling in the GIS environment and optimization focused on collection networks. The individual collection sites are grouped into the urban areas. Waste production in each urban area is estimated in dependence on available data from reports about the living environment in Prague. The locations of target collection points represented by recycling places, incineration plants and landfills are based on larger urban districts. The developed GIS project can manage various scenarios, which indicate different total costs. In spite of its actual lower total cost, the landfill will become more difficult to implement due to the increasing cost caused by limited land availability and environmental community opposition. On the other hand, the recovery/recycling will become more important because of its minimal negative interaction with the living environment. In order to minimize the costs and degradation of our living environment, spatial modeling extended by methods focused on linear programming can assist to find suitable waste management scenarios for collection of municipal solid waste in urban areas.

Keywords: municipal solid waste collection, operations research, GIS; spatial modeling.

1. INTRODUCTION

In the last hundred years, the rapid urban growth has exerted heavy pressures on resources located in the neighborhood, and resulted in serious environmental problems. A few last decades of the twentieth century saw a dramatic increase in the production of urban solid waste caused by unprecedented global levels of economic activity. Thus, urban population has been consuming land and producing wastes at an increasing rate.

Nowadays, solid waste management represents a problem of major relevance for all societies. The growing focus on environmental issues and the goal of sustainable development have moved the problem concerned on solid waste to the forefront of the public agenda. Such a consciousness has led to the development of pollution control techniques that make use of new remediation technology. Consecutively, legislation and regulations have been introduced in the local scales and in the national level to guide waste management. In addition to described management rules and techniques, strategies for sustainable waste management have emphasized the need to minimize waste production, reduce the landfill methods and increase waste recovery. Despite general agreements on the best practices for sustainable development of urban solid wastes, the goal of sustainable waste management seems far from being reached. Generally, the net waste production increases as population grows and the per capita production of waste is also increasing. Disposal in landfills is still predominant destination and the amounts of wastes being recovered are generally small. The situation is different in each particular country considering on the local costs, community opposition and restrictive environmental regulations. In general, landfill is the waste destination method that is characterized by the largest demand for land, which is a decreasing resource in urban areas [1].

Finding acceptable strategies to cope with solid waste management is becoming the inevitable task, owing to the increasing awareness of environmental issues by urban community. In the framework of European region, legislation is carried out to recommend the development of the local management plans, which accent prevention, waste reduction and recovery. In the Czech Republic, the new legislation for municipalities, the permanent increase of solid waste production and landfill shortage have encouraged the development of incineration technology and recycling programs. The methods are accommodated to take into account current economic and environmental conditions.

The aim of this work is to present the application of spatial modeling and optimization focused on municipal solid waste collection in urban regions. Spatial modeling of municipal solid waste collection represents a part of the research related with solid waste management. Principles dealing with a collection network are considered in the context of geographic information systems (GISs). GIS is able to manage spatial data containing collection networks and estimates of municipal waste production. Nowadays, optimization of collection networks can be solved with existing GIS capabilities in an efficient way, but extensions are necessary to achieve operations research functions. Generally, GIS by itself describes the spatial environment of the objects that are related to the municipal solid waste collection. Operations research optimizes the functioning of processes that are concerned on waste management. Thus, GIS and operations research need to operate together. In this case, GIS can serve as a common data framework and an analysis base for operations research. In spite of that the former GISs and operations research software evolved separately, which caused different data structures, the current GISs support a wide range of data formats, contain various case oriented extensions, and are supported by a number of programming development tools. There are a few concepts of integration both the systems on the computer platform [2, 3]. The low level integration is based on exchanging files. The GIS is mostly used to prepare the data inputs, which are then passed to another software package, and after processing, the results are returned backward to the GIS tools for display and spatial analysis together with other map layers. The higher level of integration includes sharing programming standards such as Microsoft .NET or Java that enable common running of functions through single scripts. The final step of integration can represent object oriented programming applications based on software development tools and programming libraries that cover both the GIS functionality and a selected range of the operations research methods. The higher level of integration, the more efficient running and the more experience needed to integrate all the software tools. For the aim of the presented study, the scope of the GIS integration is limited to the selected aspects of operations research focused on transportation problems.

2. Waste Management Models

Construction of any municipal solid waste collection management system in the GIS environment requires the development of a spatial data model that contains data structures of vector objects, raster layers or digital terrain models extended by attributes and topological rules. Consecutively, the data structures can be modified to support optimization methods, which are focused on particular tasks, but the nature and solution of their mathematical equations are the most critical features, rather than spatial modeling. Thus, performing these tasks within GIS requires the expression in GIS language and programming extensions.

Originally, the distinction between the spatial models in GIS and the tools for solution of operations research methods resulted in the separate development of optimization tasks in waste management information systems. Thus, waste management information systems are developed as standalone software tools and implemented models are usually based on methods for analysis of on-site collection systems of waste materials separated at the source for recovery [4, 5, 6]. The analyses can identify potential separation strategies and calculate amounts of materials collected for recovery. Finally, modeling systems offer calculation of effects of the individual separation strategies on costs and emissions. A number of time series analyses and forecasting scenarios are based on statistical methods [7]. For example, the seasonal AutoRegressive and Moving Average (sARIMA) methodology, dealing with short and medium term forecasting, can make predictions based on time series data of solid waste generated quantities.

Some case studies use methodology combining standard systems engineering with operations research methods [8, 9]. The optimization is mainly focused on the minimization of waste transportation costs in major urban areas and the comparison of waste collection strategies. The optimal number of landfills and treatment plants is used to be planned together with the optimal quantities and the characteristics of the refuse that are sent to treatment plants, to landfills and to recycling. A wealth of information on municipal solid waste management is included in the US EPA reports accessible via official web pages. In addition to the US EPA reports and guidelines, huge sources of information are published in the case studies that assist for municipal solid waste collection and optimization in specific regions.

A number of specific software tools have been developed in the last decade in co-operation with the state authorities and the local agencies. The computer programs focused on municipal solid waste collection are often developed as database applications that can manage information about amounts of municipal solid waste and its categories together with the regional demographic attributes and the economic factors. Created reports extended by basic exploratory statistical outputs are used for decision-making processes on the local and the national level. More and more actual data and software tools are accessible via the Internet.

3. Modeling and Optimization within GIS

Using GIS in the framework of waste management systems is documented by an increasing number of case studies in the latest years. Originally, GIS was used to locate potential waste disposal sites, to estimate amounts of waste in urban areas, and to find the best routes for waste collection [10]. Moreover these tasks, representation of spatial objects within a spatial domain by the data linked to the layers of points, lines, polygons and grid cells can be involved into more complex projects that enable to produce new layers based on existing data, analytical functions, numerical modeling and other external data inputs.

Recent GISs include a number of case oriented modules and extensions that are focused on partial tasks. Thus, in addition to basic modules dedicated to manage spatial data, the case oriented extensions are included to process advance tasks in spatial analysis, 3D modeling, geostatistical analysis, network optimization and spatial visualization [11, 12, 13]. For example, ArcGIS Network Analyst can assist to provide network-based spatial analyses that include routing, travel directions, closest facility, and service area analysis. The range of network analyses depends on the accessible spatial dataset that contains a range of attributes holding traffic restrictions and road conditions in dependence on specific times of the day. Thus, a number of spatio-temporal analyses can be conducted including: point-to-point routing, route directions, optimum route, service area definition and closest facility. If the functionality of GIS and its extensions are not sufficient, the GIS development tools assist to build and deploy custom applications or to extend the GIS architecture and spatial data models.

4. OPERATIONS RESEARCH

The terms operations research, management science or industrial engineering are closely related and often used synonymously. It is caused by using the interdisciplinary branch, which uses methods related to mathematical modeling, statistics, optimization, stochastic processes, graph theory, decision analysis and simulation [14]. Thus, because of the computational nature of these fields, operations research is also related to computer science. The aim of methods is usually to arrive at optimal or sub-optimal decisions in complex problems, which are concerned with searching the maxima or minima of the defined criterion. The searching is based on numerical algorithms that are represented by a large number of repeating steps, which lead to the time consuming computer processing. Thus, good or sub-optimal solutions are often sufficient [15].

The methods used to minimize the costs of municipal solid waste collection are in the rank of linear programming problems that involve the optimization of a linear objective function, subject to linear equality and inequality constrains. The most intuitive form of describing a linear programming problem consists of a linear function to be maximized:

$$c_1x_1 + c_2x_2 + ... + c_nx_n \cong \max,$$

where the constants c_1, c_2, ..., c_n are mostly the unit shipment cost and the non-negative variables x_1, x_2, ..., x_n are the quantities of the commodity that is shipped.

The other part is represented by problem constraints in the following form:

$$\begin{aligned} &a_{11}x_1 + a_{12}x_2 + ... + a_{1n}x_n \leq b_1 \\ &a_{21}x_1 + a_{22}x_2 + ... + a_{2n}x_n \leq b_2 \\ &... \\ &a_m x_1 + a_m x_2 + ... + a_{mn}x_n \leq b_m \end{aligned},$$

where the constant a_{ij} is the unit capacity of the ith source for the production of the jth commodity and the constants b_1, b_2, ..., b_m are the whole capacities for production of individual commodities. Every linear programming problem can be converted into a dual problem, which consists of a linear function to be minimized and the non-equalities in the problem constraints to be reversed. The method of linear programming implemented in the framework of municipal solid waste collection is the so-called transportation problem. It involves an assignment that considers several origins and destinations with the goal to optimize a solution at the minimal cost by minimizing transport costs with fixed origins and destinations. It considers linear transport costs based on estimates of known origins and destinations. The transportation problem can be expressed in the form of the table:

x_{11} (c_{11})	x_{12} (c_{12})	...	x_{1n} (c_{1n})	a_1
x_{21} (c_{21})	x_{22} (c_{22})	...	x_{2n} (c_{2n})	a_2
...	...		...	...
x_{m1} (c_{m1})	x_{m2} (c_{m2})	...	x_{mn} (c_{mn})	a_m
b_1	b_2	...	b_n	

In this case, there are m origins represented by resources of the municipal solid waste, and n destinations represented by landfills, incineration plants, collecting yards and recovery sites in the city districts. The quantities $a_1, a_2, ..., a_m$ of each municipal waste category are transported from each of the m origins and received in the amounts $b_1, b_2, ..., b_n$ at each of the n destinations. The constant c_{ij} is the transportation cost (in currency) per unit transported (in cubic meters or tons) from the ith resource to the jth destination place. Likewise, the variable x_{ij} represents the transportation amount (in cubic meters or tons) from the ith resource to the jth destination place. The aim of the optimization in the framework of the transportation problem is to minimize the total transport costs between a set of origins and destinations. It is assumed that the total amount of units produced and demanded is the same, which implies the equilibrium. But, any supplementary unit produced has not to be transported if there is no supplementary demand and any supplementary demand has not to be transported if there is no supplementary supply.

The transportation problem can be expressed by the mathematical model:

$$\text{Total cost} = \sum_{i=1}^{m}\sum_{j=1}^{n} c_{ij}x_{ij} \cong \min$$

$$\sum_{j=1}^{n} x_{ij} = a_i, \quad i = 1,2,...,m$$

$$\sum_{i=1}^{m} x_{ij} = b_j, \quad j = 1,2,...,n$$

$$x_{ij} \geq 0, \quad i = 1,2,...,m; j = 1,2,...,n$$

where the first equation represents the linear objective function and the rest the constraints.

Generally, the tasks in linear programming can be solved by a simplex method, but various other algorithms are in use for special cases of linear programming problems and derived methods. Their advantage resides in either a simpler calculation, or in a considerable reduction of temporarily recorded information. For example, the solution by one of the heuristic methods is based on the table introduced earlier. Step by step, the cells having the lowest transport costs are selected and the highest possible number of units in dependence on the capacity of the origin and the demand is assigned.

In each step, the corresponding transport amount must be subtracted from the related origin and the related destination for the respective row and the respective column. The iterations continue until all surpluses are used and all demands are answered. The total cost is calculated by the linear objective function. The solution based on the introduced procedure is often near the optimal solution, but other methods are needed to rich the optimal solution.

Linear programming methods, including the transportation problems, are mostly implemented as standalone software systems. Thus, in case of using GIS, a suitable level of sharing is needed to solve the complex projects containing the GIS functionality and the linear programming methods.

5. Display and Visualization of Operations Research Data in GIS

GISs mainly dedicated to manage spatial data models contain a wide range of methods that support display and visualization of transportation networks [16]. In case of linear programming, especially the transportation problems, basic GIS tools can be used to display the route network together with other thematic map layers. The case oriented extensions process advanced network analysis based on network datasets extended by connectivity rules. In addition to processing, thematic maps are generated and exported into various formats. For example, ArcGIS Network Analyst is able to carry out network analysis of spatial datasets created in the environment of ArcCatalog and ArcMap. After processing of analytical tools, the results are included into spatial data as new layers and are displayed in layout views. Then, another tool, ArcGIS ArcScene, can be used to create the visualization outputs.

6. A Case Study

As an example, the optimization of municipal solid waste collection in the urban regions of the capital Prague (area: 496 km^2; total population in December 2005: 1,181,610) is carried out to demonstrate the implementation of the transportation problem in the GIS environment. Figure 1 illustrates the central part of the city captured from satellite images. Figure 2 shows the basic map of the area represented by the individual urban areas and the street network. In case of the capital Prague, the data on municipal waste are aggregated and processed by an information system operated by the Centre for Waste Management. The overall reports and case studies are used by the Ministry of the Environment of the Czech Republic and by the Prague City Hall. Data collection, data processing and data reporting are provided in accordance with the new legislation for municipalities. The municipal waste is sorted into a few categories: paper and cardboard, colored glass, clear glass, mixed plastics, beverage boxes, bulky waste, hazardous waste, ferrous and non-ferrous metals, demolition waste, electro-technical waste, waste from care for greenery, wood waste and tires. The waste disposal is based on the kerbside collecting system or the delivery system for specific municipal waste categories. The schema of the major municipal solid waste collection subsystems is shown in Figure 3.

The sorted waste collection focused on paper and cardboard, glass and plastics is mostly provided by means of the kerbside collecting system, in which the sorted waste is turned into collecting vessels. The collection is managed by the city hall in cooperation with the operating companies. Mixed waste forms the influential part of the municipal waste in the central part of the city and on the territory of blocks of flats. The extra charges from citizens are needed to operate the disposal of the municipal mixed waste.

The mixed bulky waste is collected into large-capacity containers. The containers are allocated to the city places depending on their respective population to prevent uncontrolled dumpsite formation. The placement of containers is managed by the city districts. The bulky waste can be also turn into the collecting yards, which are operated by the city hall. The collecting yards enable to return municipal waste in a larger quantity and in a wide range of other categories. The extra waste management system is represented by the hazardous waste collection that is mostly focused on collection of toxic materials, medicines, batteries, accumulators and electronic subsystems. The waste collection is provided by a few ways: stationary and mobile collection points, pharmacies (drugs, medicines and mercury filled thermometers), schools and city offices (spent batteries and accumulators). Disposal of each municipal waste category is provided separately by the special collection equipment, transport vehicles and final treatment.

Figure 1. The capital Prague: the satellite image of the center and the surrounded areas (captured from Google Earth in September, 2007).

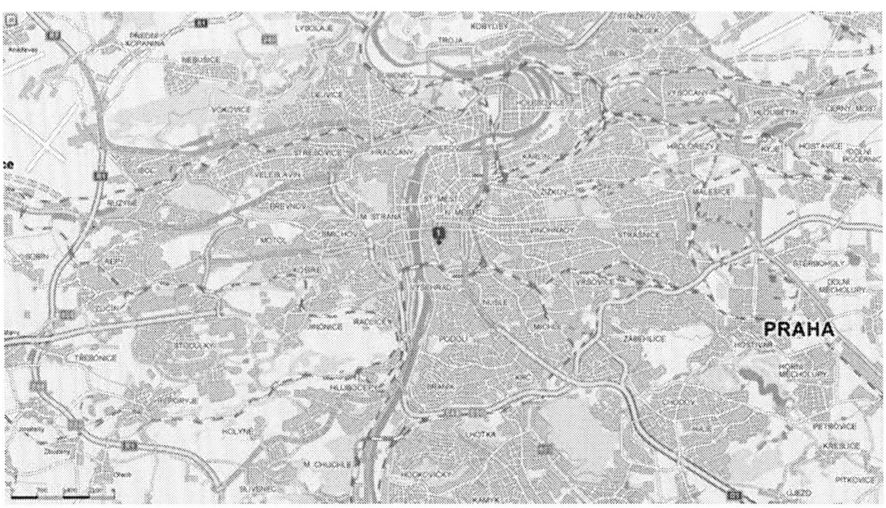

Figure 2. The capital Prague: the basic map of the center and the surrounded areas (captured from the website Mapy.cz in September, 2007).

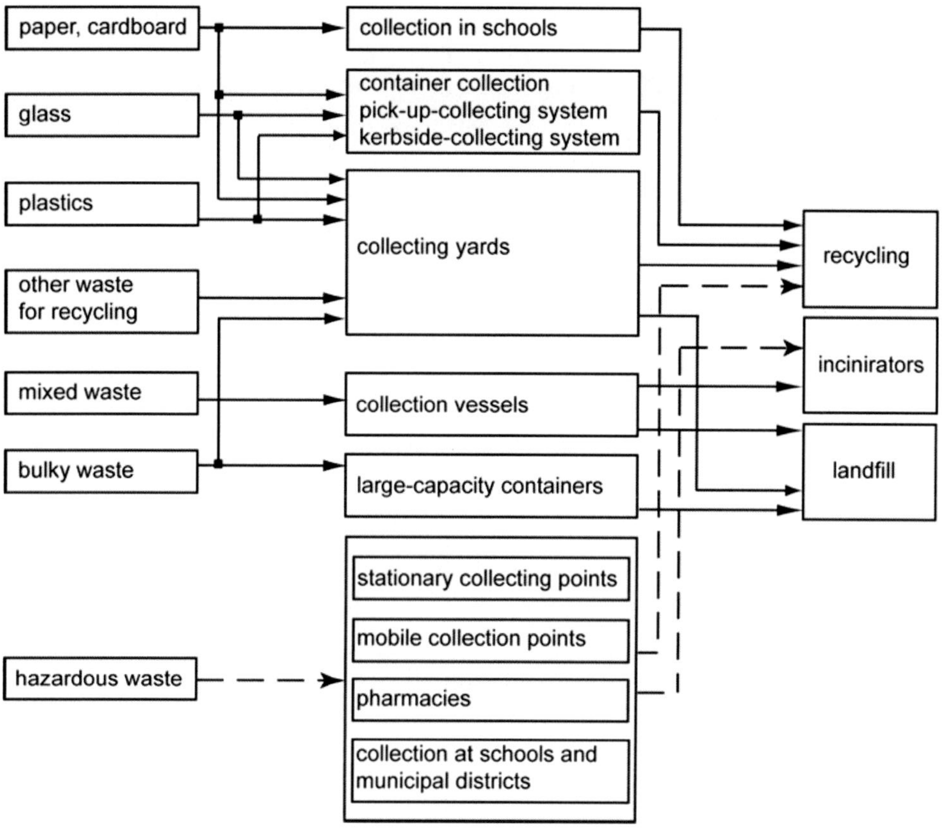

Figure 3. A basic schema of the municipal solid waste collection (based on information from the yearbooks of Prague [17] and local technical reports mostly accessible via the Internet).

6.1. Building of the Spatial Data Set in the Environment of GIS

In case of using GIS for network analysis and visualization, spatial data have to be converted into the spatial data set (ESRI's geodatabase or a specified set of files). The spatial dataset contains geographic locations of spatial objects and their topological rules, which are necessary for network analysis. To demonstrate the optimization of municipal solid waste collection in a simple way, the collection places for each category are reduced into the central points representing individual urban areas as the resources. The resources are divided into a few basic categories: paper, mixed plastics, mixed glass and mixed waste. Other categories like hazardous waste, bulky waste and metals are not included into the set, because their collection management is different. The set of target points is represented by recycling places, incineration plants and landfills. In order to solve the transportation problem and present the principles of waste collection in the GIS environment, the number of target points is reduced to simplify the structure. As an example, the GIS project is developed to demonstrate spatial data management, network analysis and sharing operations research data. The waste resources and waste targets are included into the GIS map layers and illustrated together with the city road network in Figure 4.

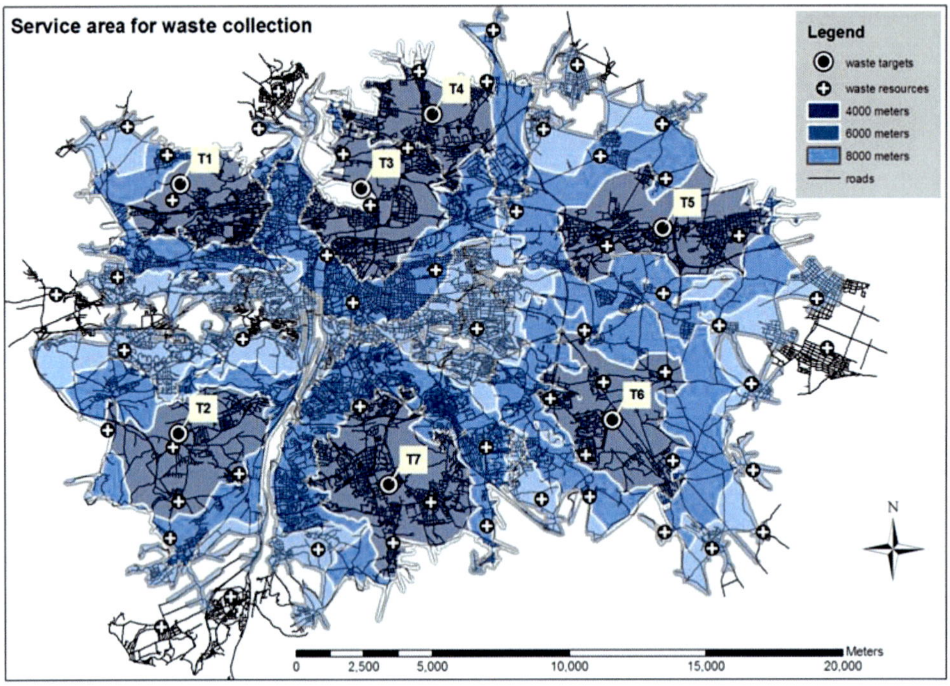

Figure 4. The schema of waste resources and waste targets extended by service areas.

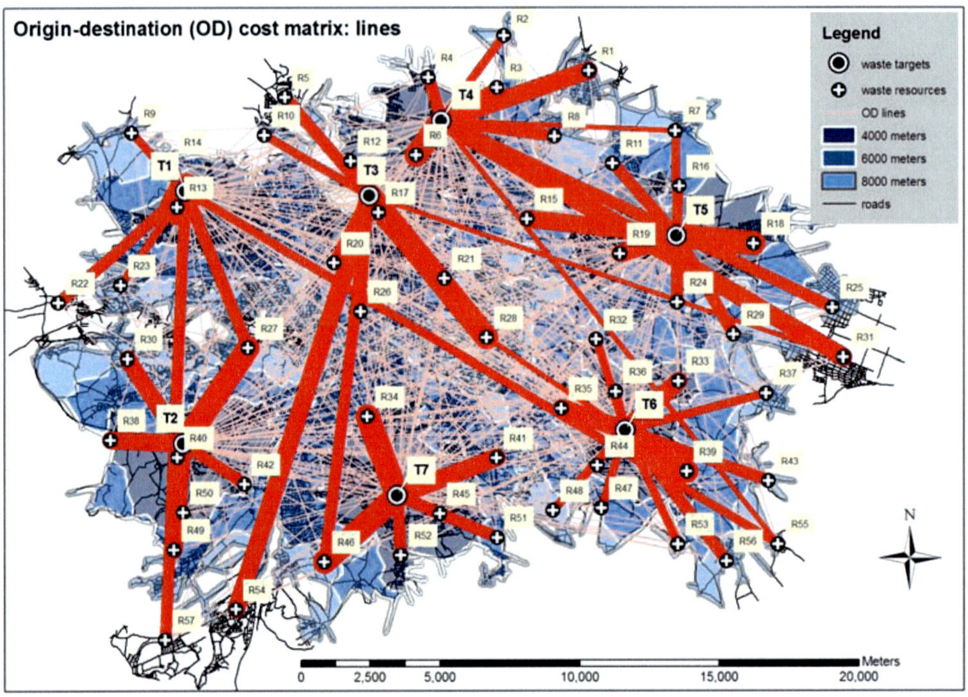

Figure 5. The schema of waste resources and waste targets extended by OD lines.

The GIS extensions can solve a number of network analyses based on finding the best route among a set of resources or finding the closest target points. In case of this example, spatial analysis is used for calculating the service area and creating the Origin-Destination (OD) cost matrix. The service areas around the target points for selected road distances (4000, 6000, 8000 meters) are displayed in Figure 4. The GIS network extensions (NetworkAnalyst in ArcGIS) is used to create a series of polygons representing the distance that can be reached from target points within a specified road distance. The OD cost matrix is represented by a series of the connection lines in Figure 5. The matrix is used as the input for the transportation problem. Each connection line includes a number of attributes. The road distance between the resource point and the target point is used to estimate the transportation cost per unit transported from the ith resource to the jth target. The estimated transportation costs together with the estimated quantities of each municipal waste category at the resources are transferred into the numerical module for solution of the transportation problem. After the calculation, the optimized transportation amounts between the resources and targets are backward transferred into the GIS project. The solution based on optimization of the total transport costs between the set of resources and the set of targets is illustrated in Figure 5.

The width of the OD lines represents the relative transport amounts among the individual pairs. The thin connection lines on the background show the whole set of pairs, which, in this case, do not serve for transportation.

6.2. Sharing Data between Operations Research Tools and GIS

The presented case study demonstrates calculation of the transport problem by sharing data between GIS and the specialized programming module. In case of the selected city, the datasets contain 57 resources and 7 targets for each municipal waste category. The estimates of attributes in the table (c_{ij}; R_i; T_j) are based on GIS spatial data (road distance, size of each area). The transportation problem is expressed by the linear objective function:

$$\text{Total cost} = \sum_{i=1}^{57}\sum_{j=1}^{7} c_{ij}x_{ij} \cong \min$$

and by the constraints:

$$\sum_{j=1}^{57} x_{ij} = a_i, \quad i = 1, 2, ..., 57$$

$$\sum_{i=1}^{7} x_{ij} = b_j, \quad j = 1, 2, ..., 7$$

$$x_{ij} \geq 0, \quad i = 1, 2, ..., 57; j = 1, 2, ..., 7$$

The output variables (x_{ij}) represent the transportation amounts. The solution based on heuristic methods is expressed in the table:

Waste category: (paper, mixed plastics, mixed glass and mixed waste)							
$x_{1,1}$ ($c_{1,1}$)	$x_{1,2}$ ($c_{1,2}$)	$x_{1,3}$ ($c_{1,3}$)	$x_{1,4}$ ($c_{1,4}$)	$x_{1,5}$ ($c_{1,5}$)	$x_{1,6}$ ($c_{1,6}$)	$x_{1,7}$ ($c_{1,7}$)	R_1
$x_{2,1}$ ($c_{2,1}$)	$x_{2,2}$ ($c_{2,2}$)	$x_{2,3}$ ($c_{2,3}$)	$x_{2,4}$ ($c_{2,4}$)	$x_{2,5}$ ($c_{2,5}$)	$x_{2,6}$ ($c_{2,6}$)	$x_{2,7}$ ($c_{2,7}$)	R_2
...	...	...	...	...	...	...	...
$x_{57,1}$ ($c_{57,1}$)	$x_{57,2}$ ($c_{57,2}$)	$x_{57,3}$ ($c_{57,3}$)	$x_{57,4}$ ($c_{57,4}$)	$x_{57,5}$ ($c_{57,5}$)	$x_{57,6}$ ($c_{57,6}$)	$x_{57,7}$ ($c_{57,7}$)	R_{57}
T_1	T_2	T_3	T_4	T_5	T_6	T_7	

CONCLUSION

The synthesis of GIS and operations research, likewise dynamic modeling [18], in the framework of waste management focused on municipal solid waste collection has the potential to create a new base for optimization tools that assist to solve waste treatment and waste disposal in a more efficient way. Utilizing GIS and development operations research tools on it is making cost effective and highly efficient systems, which can be implemented into urban planning and decision-making processes. Using GISs for brings possibilities of using standard spatial data formats, implemented analytical methods and visualization tools. The outputs integrated into the map layers and managed by GIS servers offer new ways of sharing information on the local level of the city regions. The operation research methods have been using for a few decades. Despite time consuming numerical algorithms for solution of the optimization tasks, the progress in computer technology and new GIS software can be extended by the case oriented modules focused on operations research methods including linear programming.

The attached case study illustrates the potential possibilities for optimization of waste management tasks. In addition to the standard solution of the transportation problem, the setting of the higher transportation costs between the resources and the targets represented by disposal in landfills or incineration plants enable to carry out the new scenarios, which are more sustainable and protect our living environment. Thus, the extra costs can be used for recycling and building of the new recovery sites. As an example, only one category of municipal solid waste is selected to display the service areas and the OD lines in the previous figures, because the vehicles use the same road network for other waste categories. However, the transportation amounts between the set of resources and the set of targets depend on each category.

The solution of optimization tasks for each category of the municipal solid waste offers sharing of the data outputs together with other map layers in the GIS environment, which help in urban planning and environmental protection. New scenarios can be created by modification of data inputs into transportations problem in order to find the suitable solution.

ACKNOWLEDGEMENTS

The spatio-temporal data and operations research methods have been processed by ArcGIS and programming software tools in the GIS Laboratory, Faculty of Natural Science, Charles University in Prague.

REFERENCES

[1] Leao, S., Bishop, I. and Evans, D. (2001). Assessing the demand of solid waste disposal in urban region by urban dynamics modelling in a GIS environment. *Resources, Conservation and Recycling 33*, 289-313.

[2] Maidment, D.R. (1996). Environmental Modeling within GIS. In M.F. Goodchild, L.T. Steyaert, B.O. Parks (Ed.), *GIS and Environmental Modeling: Progress and Research Issues* (pp. 315-323). Fort Collins: GIS World, Inc.

[3] Fedra, K. (1996). Distributed Models and Embedded GIS: Integration Strategies and Case Studies. In M.F. Goodchild, L.T. Steyaert, B.O. Parks (Ed.), *GIS and Environmental Modeling: Progress and Research Issues* (pp. 413-417). Fort Collins: GIS World, Inc.

[4] Tanskanen, J-H. (2000). Strategic planning of municipal solid waste management. *Resources, Conservation and Recycling 30*, 111-133.

[5] Choe, C. (1999). An Economic Analysis of Household Waste Management. *Journal of Environmental Economics and Management 38*, 234-246.

[6] Barlishen, K.D. and Baetz, B.W. (1996). Development of a Decision Support System for Municipal Solid Waste Management Systems Planning. *Waste Managementand Research 14*, 71-86.

[7] Navarro-Esbrí, J., Diamadopoulos, E. and Ginestar, D. (2002). Time series analysis and forecasting techniques for municipal solid waste management. *Resources, Conservation and Recycling 30*, 201-214.

[8] Kulcar, T. (1996). Optimizing solid waste collection in Brussels. *European Journal of Operational Research 90*, 71-77.

[9] Fiorucci, P., Minciardi, R., Robba, M., Sacile, R. (2003). Solid waste management in urban areas Development and application of a decision support system. *Resources, Conservation and Recycling 37*, 301-328.

[10] Muttiah, R.S., Engel, B.A., Jones, D.D. (1996). Waste Disposal Site Selection Using GIS-based Simulated Annealing. *Computers and Geosciences 22*, 1013-1017.

[11] MacDonald, A. (2001). *Building a Geodatabase*. Redlands, California: ESRI Press.

[12] Maguire, D.J., Batty, M., and Goodchild, M.F. (2005). *GIS, Spatial Analysis and Modeling*. Redlands, California: ESRI Press.

[13] Mitchell, A. (2005). *The ESRI Guide to GIS Analysis. Volume 2: Spatial Measurements and Statistics*. Redlands, California: ESRI Press.

[14] Dantzig, G.B. (1963). *Linear Programming and Extensions*. Princeton, Princeton University Press.

[15] Fletcher, R. (1987). *Practical Methods of Optimization*. New York, Wiley.

[16] Maguire, D.J., Batty, M., and Goodchild, M.F. (2005). *GIS, Spatial Analysis and Modeling*. Redlands, California: ESRI Press.

[17] Solc, J. and Pospisilova, J. (2007). Prague-Environment 2006. Prague, J.I.M. Liberec.

[18] Matejicek, L. (2006). Modeling of Water Pollution in River Basin with GIS. In A.R. Burk (Ed.), *Focus on Ecology Research* (pp. 249-270). New York, Nova Science Publishers, Inc.

In: Solid Waste: New Research
Editors: A. S. Martin and J. D. Suarez
SBN: 978-1-60456-809-7

Chapter 7

BIOLOGICAL TREATMENT OF ORGANIC SOLID WASTE IN ANAEROBIC REACTORS

Wilton Silva Lopes[1,*], Valderi Duarte Leite[2,†], Shiva Prasad[3,‡] and José Tavares de Sousa[4,§]
[1,2,4]Department of Chemistry, Universidade Estadual da Paraíba, CEP 58100-001. Campina Grande, Paraíba, Brazil
[3]Department of Chemical Engineering, Universidade Federal de Campina Grande, CEP 58109-970. Campina Grande, Paraíba, Brazil

ABSTRACT

The purpose of this work is to evaluate the influence of the principal intervening factors in anaerobic treatment process of organic solid waste. The results obtained from the research work on biological treatment of organic solid waste in anaerobic reactors which has been realized since 1997 by the Environmental Sanitation Group of Universidade Estadual da Paraíba, northeast of Brazil will be presented. During these years various operational parameters such as: size and configuration of the reactor, moisture amount, type and amount of inoculum, and others have been studied. Treatments with high and low concentrations of solids were realized and the results obtained showed that the removal efficiency of the applied mass to the reactor and the moisture amount are directly proportional. During the anaerobic treatment with high concentration of solids, one problem that can be verified was the high concentration of volatile fatty acids and chemical oxygen demand in the liquid leachate, which tends to lower the pH down to the levels between 2.0 to 4.0, thus inhibiting the methanogenic microorganisms. During the process of anaerobic treatment with high concentration of solids the biogas produced were almost totally composed of CO_2. When the process attains the dynamic equilibrium the rate of CH_4 production attains the maximum, then it decreases continuously through the end of the reactor operation. In the case of anaerobic

* E-mail: wiltonuepb@yahoo.com.br
† E-mail: valderileite@uol.com.br
‡ E-mail: prasad@deq.ufcg.edu.br
§ E-mail: jtdes@uol.com.br

treatment with low concentration of solids, the biogas production is practically constant after attaining the equilibrium in the reactor. Gas chromatographic analysis of the biogas showed that the amount of methane went as high as 70%, depending upon the applied operational parameters. Bovine rumen, UASB sludge and leachate were used as inoculum, and it was verified that in general the biostabilization time of the waste was inversely proportional to the percentage of inoculum used, and independent of the inoculum type. Scanning electron microscopic and fluorescence optical microscopic analyses of the samples from the reactors showed predominance of *Methanosarcina sp.* kind of microorganisms, which explains the high amount of methane in the biogas. In general, it can be concluded that the various intervening factors existing in the organic solid waste treatment process in anaerobic reactors have made difficult the application of this type of process at real scale. Thus, one should be very careful while projecting this type of treatment system.

1. INTRODUCTION

1.1. Organic Solid Waste

Organic solid wastes constitute a large part of municipal solid waste, making up about 50wt% of agro-industrial solid waste, rural and agricultural solid waste, some types of industrial waste and residential sanitary sewage treatment sludge. In Latin America, most of the organic waste present in municipal solid waste is disposed of in sanitary land-fills, which do not possess any basic infra-structure required by environmental and sanitary engineering. It generates leachate with high concentration of COD, TKN and heavy metals, thus polluting the environment, besides the emission of biogas (CO_2, CH_4, H_2S, etc.) directly to the atmosphere.

Organic solid waste is classified in FOFMSW (fermentable organic fraction of municipal solid waste); ASW (agricultural or rural solid waste); AISW (agro-industrial solid waste) and SSS (sanitary sewage sludge). In these fractions, the more representative is FOFMSW, since it represents on an average about 50wt% of municipal solid waste.

Residential Organic Waste: is basically composed of leftovers of food, fruits, vegetables discarded during kitchen and dining activities, foliage collected through sweeping the courtyard and others.

Agricultural or Rural Organic Waste: is all the solid waste produced by agricultural activities. In general, it is composed of leftovers from cultivation, different types of vegetation, animal excrements and other organic deposits.

Agro-industrial Organic Waste: includes all the organic waste originated from different agro-industrial activities, such as the activities of processing cereals, milk and its derivatives, sweets, meat and derivatives, oil, etc.

Sludge: sludge generated only from residential sanitary sewage treatment activities will be focused here. Therefore, the sanitary sludge can contain a percentage of organic material, which in general depends upon the type of treatment applied to sewage, on the collecting system of sewage and on the project of the respective sewage treatment plant. Depending on the treatment system, the sludge can be aerobic or anaerobic, presenting peculiarities in their chemical and physical proprieties. The sludge produced is estimated to be about $54gTS.hab^{-1}.day^{-1}$, which corresponds to 50% of per capita production of residential organic solid waste.

1.2. Principal Processes of Biostabilization of Organic Solids

Organic solid waste can be biostabilized by aerobic or by anaerobic methods. By the aerobic route the well-known and amply diffused process is composting whose objective is to produce compost for applying in agriculture. However, in Brazil only 2.8% of fermentable organic material is being treated in composting plant. In north and northeast regions of the country the data demonstrate that the utilization of composting plant for treatment of fermentable organic fraction of the municipal solid waste is almost inexistent. All the organic material is thrown in open land, which together with other types of materials present in the municipal solid waste produces harmful effect to the environment. In northeast region of the country, for example, about 21,194 tons of fermentable organic material are produced daily, which could produce approximately 18,000 tons per day of organic compost, with an ample space for utilization, principally in rural zone populated by small and medium level producers.

Fermentable organic material of municipal solid waste collected at national level contains about 65% TVS, which can significantly contribute for application of this material in aerobic or anaerobic process.

Aerobic composting can be considered as a biological process carried out by bacteria, fungi and actinomycetes, in presence of free oxygen with substrate containing about 50% moisture, producing the biostabilized organic compost and carbon dioxide, water and mineral substances as byproducts.

Composting can be done in presence or in absence of free oxygen, or it can also be a mixed composting. In Brazil, still the principal method used is aerobic in open atmosphere with periodic upturning, for controlling the temperature and oxygenation of the waste mass. The main factors that should be controlled in the composting process are: temperature, pH, moisture, aeration, granulometry and C/N ratio. The principal advantages of the composting process are: possibility of correction of soil structure, increase in water absorption capacity of soil, improvement in microbial activity of soil, increase in aeration and macro and micro nutrients of soil, favoring still the increase in soil stability with correct level of acidity.

2. Anaerobic Biostabilization Process for Organic Solid Waste: Fundamentals

Anaerobic processes were used by ancient Roman people and, ever since have been used for diverse purposes, principally for treating waste containing high organic loads.

Anaerobic digestion can be considered one of the oldest technologies for stabilization of waste water. It has been applied since the end of XIX century, principally for the treatment of domestic sewage in septic tanks and for the treatment of sewage sludge (van Lier *et al.*, 2001).

In Brazil, anaerobic process have been used for biostabilization of sanitary sewage sludge and liquid effluents with high biochemical demand of oxygen and, in some specific cases, solid waste of rural origin.

According to Barlaz and Ham (1989), anaerobic biostabilization process of organic solid waste, obeys the same metabolic routes already established for anaerobic biostabilization of sludge.

The bacterial metabolism consists of utilization by bacteria of organic material as energy source for diverse activities inherent to realization of the process (van Haandel and Lettinga, 1992). In general, the anaerobic biostabilization process starts when the volatile suspended solids fed to the system are hydrolyzed, producing soluble COD. The soluble COD serves as food for the bacteria and, consequently, the cellular synthesis, giving origin to a relatively dense microbial biomass. In endogenous respiration conditions, a part of the biomass returns back by cell lysis for producing soluble COD, and the rest of it forms an inert mass (Pavlostathis, 1988).

Anaerobic biodigestion is a biochemical process that occurs in absence of free molecular oxygen, in which different species of microorganisms interact among themselves in symbiotic form to convert complex organic compounds in CH_4, inorganic compounds such as CO_2, N_2, NH_3, H_2S and traces of other gases and organic acids of low molecular weight.

Application of anaerobic digestion for sewage treatment has increased during the last decades with the objective of developing new advanced systems with high biomass retention. In relation to treatment of solids, the anaerobic biodigestion process has been studied since 1960's, trying to develop technologies to improve the energy recovery and reduction of the volume and organic mass (Peres *et al.*, 1991).

According to Lay *et al.* (1998), anaerobic digestion of organic solid waste, especially of the fermentable organic fraction of municipal solid waste, is of great importance in management of solid waste. Various types of reactors have been developed for treatment of organic fraction of municipal solid waste, which include the continuous and the batch reactors. But, the solid waste treatment still does not possess a large number of options of biological reactors for stabilization of the organic material.

The anaerobic treatment process appears as a promising alternative for the organic solid waste due to high production rate of biogas. However, the anaerobic processes employed in solid waste treatment still do not constitute a very well diffused practice due to lack of configuration of treatment systems and, mainly due to the time required for biostabilization of solid waste anaerobically, which is quite long when compared with that of the aerobic processes.

Use of some inoculums has shown satisfactory results in decreasing the biostabilization time of organic solid waste. The inoculums generally used are sludge from sewage treatment stations or some material from animal origin, such as bovine manure and others.

According to Barlaz and Ham (1989), when treating the biostabilization process of fermentable organic fraction of municipal solid waste, cellulose together with hemicellulose are responsible for approximately 91% of the energetic potential of the process. Besides this, the author reported that the mineralization of cellulose and hemicellulose attained to 71 and 77% respectively after 111 days of functioning of the experimental system.

Anaerobic digestion process is influenced by several factors, such as temperature, applied organic load, presence of any toxic material and others. At high temperatures, the biological reactions occur with higher velocity, resulting possibly in a higher efficiency of the process. In general, the anaerobic process can be developed at mesophilic (30 to 45^0C) or thermophilic (45 to 60^0C) level. Regarding the organic load, various types of reactors and substrates have already been tested for different loads. In comparison to other types of treatments, the anaerobic process responds satisfactorily well to the load fluctuations, principally when the reactors are already in operation.

3. Kinetics of Anaerobic Biostabilization Process of Organic Solid Waste

Kinetic study of a phenomenon or a process means to study the evolution of the process in the time through quantification of certain parameters that define adequately this production. Like this, the kinetic study of a certain fermentation process means to study the alterations that occur in the process in a determined time. The measurements of concentration of microorganisms present (X), concentration of the substrate that limit the process (S) and concentration of the product (P) of interest are generally made (Schmidell Netto, 1999).

The microbial growth can be considered as a set of biochemical chain reactions, which synthesize the constituents of microbial biomass obtained at the end of the operation (Leveau and Bouix, 1985).

The standard growth profile of microorganisms in a closed system divided in four phases is shown in Figure 1.

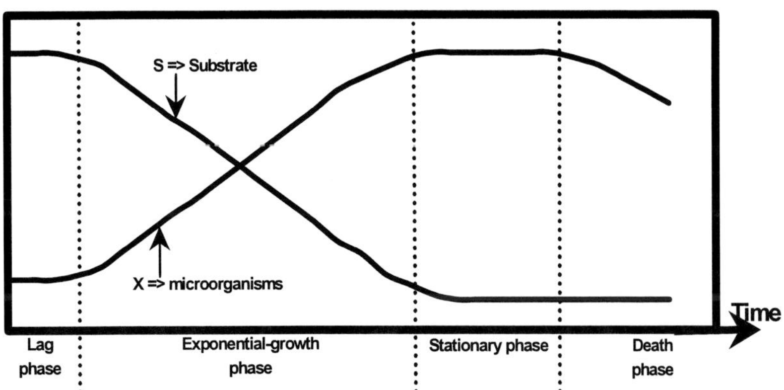

Figure 1. Bacterial growth profile. Adapted from: Bitton, 1999; Metcalf and Eddy, 2003

Lag phase: this is the phase of adaptation by the organism in new medium for starting its division when an inoculum is added to a culture medium;

Exponential-growth phase: during this period, the cells divide at the rate determined by their generation time and by their ability to the feeding process;

Stationary phase: when the cell population remains stationary. The reasons for this phenomenon are: that the cells consumed all the substrate or nutrients for growth or that the growth of new cells is equilibrated with the death of old cells;

Death phase: during this phase, the bacterial death rate exceeds the production rate of new cells. The death rate is usually a function coming from the population viability and environmental characteristics. In some cases, the logarithmic decline phase is inverse of the logarithmical growth phase (Bitton, 1999; Metcalf and Eddy, 2003).

In biological processes of waste treatment the principal phase in the bacterial growth profile is that of the exponential-growth, since the degradation rate of the organic material is a function of the number of the microorganisms present, thus one expects that the maximum concentration of the bacterial mass is attained. And the least interesting phase is that of the adaptation because of the insignificant growth rate.

Because of increase in microorganisms the substrate concentration (S) tends to decrease with the lapse of time. The substrate consumption velocity can be obtained by tangent of the curve corresponding to the exponential-growth phase in Figure 4. Therefore the substrate consumption velocity can be represented by equation 1:

$$r_S = -\frac{dS}{dt} \tag{1}$$

And specific velocity of substrate consumption by equation 2:

$$r_S = -\frac{dS}{dt} \Rightarrow \frac{r_S}{S} = -\frac{dS}{dt}\cdot\frac{1}{S} \Rightarrow k = -\frac{1}{S}\cdot\frac{dS}{dt} \tag{2}$$

by integrating equation 2, one gets:

$$S = S_0 e^{-k(t-t_0)} \tag{3}$$

where: S = substrate concentration at time t ($mg.L^{-1}$);
S_0 = initial concentration of the substrate ($mg.L^{-1}$);
k = biodegradability constant of the substrate.

Biodecomposition of organic substrate in anaerobic batch reactor is characterized by a slow phase, which is the necessary period for adaptation and beginning of multiplication of bacterial mass, besides a fast decomposition phase, whose bacterial growth is the maximum. Therefore, the substrate biodegradation rate can be described by exponential models (Leite *et al*, 2001).

Many authors describe the hydrolysis behavior of organic material complying with the first order kinetics by taking the substrate biodegradability as base at constant pH and temperature, as described in equation 3 (Sanders *et al.*, 2000; Lopes *et al.*, 2003).

In most of the works published, related to biological treatment of organic solid waste the parameter used as reference for evaluation of performance of the process is the efficiency of reduction of the substrate mass applied or reduction of the total volatile solids.

Biodegradability constant values can be used for evaluation of performance of the process. As it is the case of exponential equations, therefore higher the magnitude of k better will be performance of the process.

Values of k are characteristic for each type of substrate and there exist various factors which can influence in degradation rate of the substrate, principally, when the substrate used is a solid. In case of the organic solid waste, various factors can be mentioned such as: concentration of solids, granulometry, chemical composition, presence and amount of inoculum. The exponential equations which were adjusted with the data obtained by Lopes *et al.* (2004) during the anaerobic treatment of fermentable organic fraction of the municipal solid waste inoculated with bovine rumen in different proportions are presented in Table 5. The data show that with increase of inoculum amount there is an increase in magnitude of values of the constant.

Table 1. Adjusted exponential equations

% Inoculum	TVS	TKN	S
0	$TVS=1431.e^{-0.001028.t}$	$TKN=43.e^{-0.001852.t}$	$S=5.3.e^{-0.001232.t}$
5	$TVS=1322.e^{-0.001536.t}$	$TKN=42.e^{-0.002999.t}$	$S=4.9.e^{-0.001932.t}$
10	$TVS=1387.e^{-0.001453.t}$	$TKN=43.e^{-0.003140.t}$	$S=5.1.e^{-0.001850.t}$
15	$TVS=1327.e^{-0.002105.t}$	$TKN=40.e^{-0.003140.t}$	$S=5.4.e^{-0.003361.t}$

The time t is expressed in days, TVS (total volatile solids), TKN (total nitrogen) and S (sulfur) in grams.

From: Lopes *et al.*, 2004.

Presently there exist evidences that the reactor size has influence on the values of the constant. Borzacconi *et al.* (1997), obtained a value for hydrolysis constant $1,6x10^{-3}day^{-1}$ for a reactor of 200 liters and a value of $0.8x10^{-3}$ day^{-1} for a reactor of 10 m^3, both treating anaerobically the fermentable organic fraction of the municipal solid waste. This shows that during the anaerobic biostabilization of municipal solid waste reactor size influences in performance of the process, seeing that the greater the hydrolysis constant smaller the time required for biostabilization of material in the reactor.

Christ *et al.* (2000), when studied the hydrolysis constant of anaerobic treatment process of solid waste with low concentration of solids (between 5 and 10% of suspended solids) at 55°C, obtained the following results:

Table 2. Hydrolysis constant values for different fractions of organic waste

Process	$k_{minimum}$ (day^{-1})	$k_{maximum}$ (day^{-1})	$k_{maximum}$ / $k_{minimum}$
Hydrolysis of lipids	0.005	0.010	2
Hydrolysis of proteins	0.015	0.075	5
Hydrolysis of carbohydrates	0.025	0.200	8

Adapted from: Christ *et al.*, (2000)

The duplication time of bacteria in anaerobic systems is very divergent, and this contributes for that the stationary phase in this type of system delays to be attained. Some data related to duplication time of groups of anaerobic bacteria responsible for decomposition of sugars at 35°C are presented in Table 3.

Table 3. Duplication time for mesophilic anaerobic microorganisms responsible for decomposition of sugars

Microorganisms	Duplication time
Fermentatives that produce acid	30 min.
Methanogenics that consume H_2 or formiate	6 h
Acetogenics that consume butirate	1.4 days
Acetogenics that consume propionate	2.5 days
Metanogenics that consume acetate	2.6 days

Adapted from: Viñas (1994); Nascimento (1996)

Analyzing Table 3 it can be observed that the methanogenic microorganisms have a low growth rate when compared to the other microbial groups involved in the process. This results in that the methanogenic microorganisms generally limit the velocity of anaerobic digestion process.

In relation to anaerobic treatment of fermentable organic fraction of municipal solid waste, the idea that methanogenic microorganisms are the limiting factor of the process is currently much discussed and some workers (Christ *et al*, 2000; van Lier *et al*, 2001; Bouallagui *et al*, 2005) believe that in anaerobic treatment of this type of waste the phase that limits the process is hydrolytic.

4. Microbiology of the Anaerobic Treatment Process of Organic Solid Waste

In anaerobic biodigestion systems the degradation of organic material involves activity of obligately anaerobic and facultative microorganisms, whose species belong to the group of hydrolytic, fermentative acidogenic, acetogenic and methanogenic (Bitton, 1999). The microorganisms of anaerobic biodigestion have a high degree of metabolic specialization. Therefore the efficiency of anaerobic process depends on positive interactions between diverse bacterial species, with different degradation capacities. The intermediary metabolites of a group of bacteria can serve as nutrients for growth of other species. Thus, the occurrence of various reactions of degradation of the organic compounds and their dependence on presence of the hydrogen formed in the system is observed. The scheme of the reactions that occur during degradation of organic solid waste in anaerobic systems is presented in Figure 2 (Gujer and Zender, 1983; Stronach *et al.*, 1986; Bouallagui *et al.*, 2005).

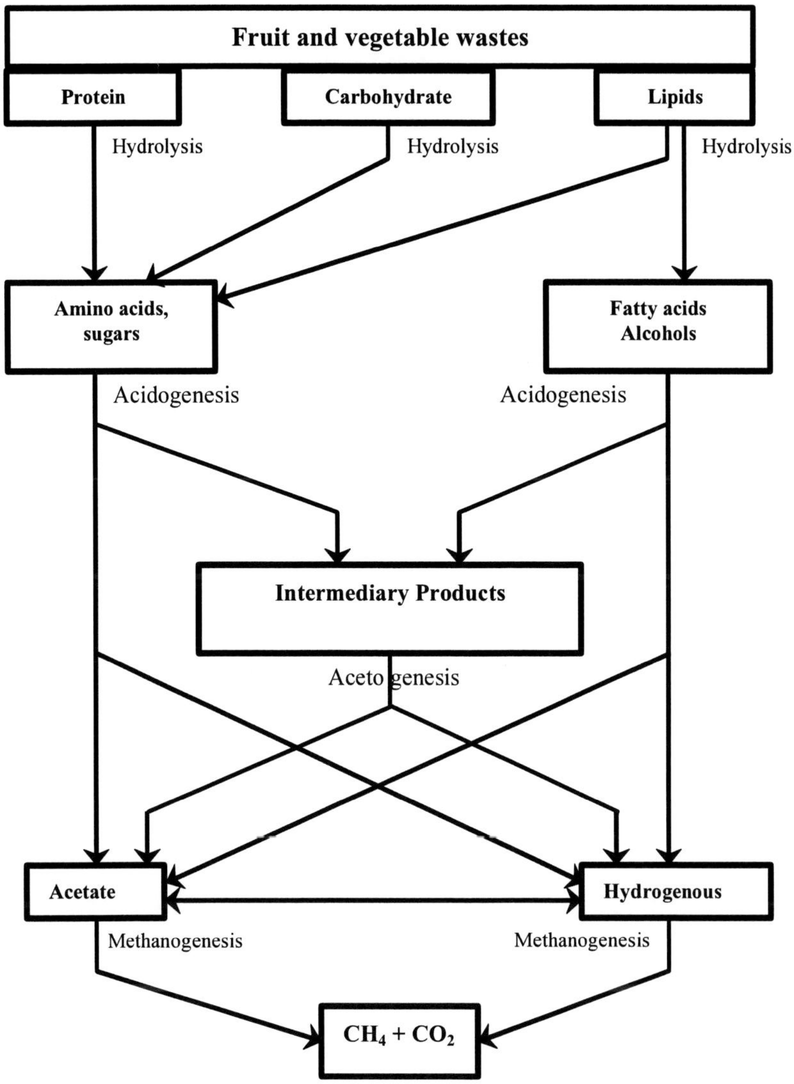

Figure 2. Scheme of degradation of organic solid waste in anaerobic systems. Adapted form: Gujer and Zender, 1983; Stronach *et al.*, 1986; Bouallagui *et al.*, 2005

The principal groups of microorganisms that act in anaerobic process are:

- *fermentative microorganisms* – are facultative anaerobic microorganisms, responsible for the first two stages of the process: hydrolysis and acidogenesis;
- *acetogenic microorganisms producers of hydrogen* – are bacteria that catabolize organic acids of chains longer than that of acetic acid, alcohols and some aromatic compounds to hydrogen, carbon dioxide and acetic acid. There occurs a syntrophic relation between these microorganisms and the producers of methane and consumers of hydrogen, for equilibrating the process in terms of production and consumption of hydrogen;

- *homoacetogenic microorganisms* – are microorganisms that metabolize compounds of only one carbon atom, besides utilizing hydrogen for producing acetate;
- *methanogenic microorganisms* – this group of microorganism are divided in function of the substrate utilized in:
  - acetoclastic methanogenics – represent a few species that convert acetate in methane, and are predominant microorganisms since they are responsible for about 70% of the methane produced. This group is the most versatile among the methanogenics, since they possess species capable of utilizing hydrogen and methylamines too;
  - hydrogenotrophic methanogenics – practically all the species of methanogenics belong to this group that produce methane from hydrogen and carbon dioxide.

Different types of microorganisms which interact among themselves in anaerobic processes for converting complex organic compounds in biogas are presented in Table 4.

Table 4. Types of microorganisms intervening in anaerobic processes

Types	Microorganism Species
Fermentative microorganisms	Clostrídios, Acetivibrio cellulolyticus, Bacteroides succinogenes, Butyrivibrio fibrisolvens, Eubacterium cellulosolvens, Bacillus sp, Selenomonas sp, Megasphaera sp, Lachnospira multiparus, Peptococcus anaerobicus, Bifidobacterium sp, Staphylococcus sp
Acetogenic microorganisms	Syntrophomonas wolinii, S. wolfei, Syntrophus buswellii, Clostridium bryantii, Acetobacterium woddii
Acetoclastic methanogenic microorganisms	Methanosarcina sp e Methanothrix sp
Hydrogenotrophic methanogenic microorganisms	Methanobacterium sp, Methanobrevibacter sp, Methanospirillum sp

From: Adapted from Chernicharo (1997) and Vazoller (1999)

Bryant (1974) was the first person who classified the methanogenic microorganisms by defining the Methanobacteriaceae family as constituted of three species: Methanobacterium, Methanosarcina and Methanococcus. The methanogenic microorganisms are classified in Archaebacteria kingdom and differ from Eubacteria kingdom in their metabolic and nutritional aspects (Oremland, 1988).

Philogenetically the methanogenic bacteria are considered as the most primitive group of live microorganisms present on the earth. They are considered as obligately anaerobics and use only some types of substrates. They develop favorably in reduced medium with redox potential of about -300mV (Zehnder and Stumm, 1988).

The methanobacteria grow in environment with mesophilic (15 to 45°C) and thermophilic (55 to 65°C) temperatures. The pH interval for the production of methane is situated in the range 6.8 o 7. 2. However, in some instances the production of methane from organic waste has been verified at pH level lower than 6.0. They show necessity of normal nutrients and most of the species require only mineral salts, carbon dioxide, ammonia and sulfide.

In this work, the microscopic tests were performed on biofilm formed on filtering bed of the reactors. The filtering bed was a device composed of sand of different granulometry and was used for retaining small and large sized material, thus avoiding the obstruction of the leachate at collection point. The samples were collected from the sediment formed on upper surface of the filtering bed and along its depth, where the smaller grains of sand were covered by a thick layer of biomass. The scanning electron microscopic tests were performed by using scanning electron micrograph, Jeol, JSM 25 SII, whereas the common optical fluorescence and phase contrast tests were done by Olympus BHT Microscope (ocular 10, objective 100 and zoom 1.25). The basic procedure employed on the samples for the scanning electron micrograph tests were: cleaning, fixing, dehydration, mounting, metallization and recording the sample readings. The tests were performed by using the method given by Santos (1992). The fluorescence microscopic and phase contrast tests were performed by the following procedure: dilution, fixing and then reading the samples.

The results of scanning electron micrograph analysis of the material collected from filtering bed of some reactors during the experimental work are shown in Figures 3, 4 and 5.

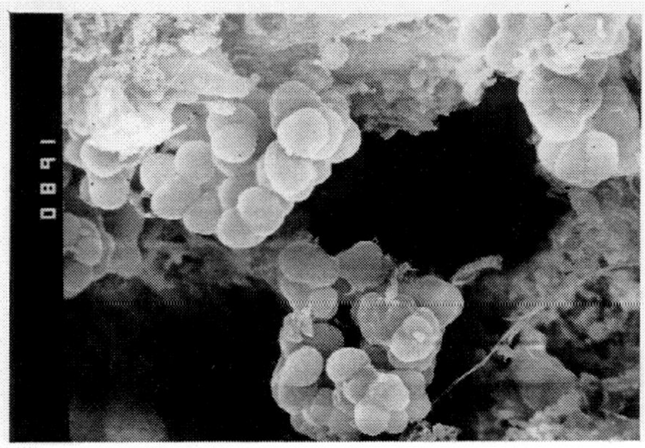

Figure 3. Scanning electron micrograph. Sample (3000x; scale 10 m)

Figure 4. Scanning electron micrograph (1500x; scale 10 m)

Figure 5. Scanning electron micrograph. (3000x; scale 10 m)

Results of common optical fluorescence and phase contrast tests, performed on the material collected from filtering bed of one of the reactors used, on a sample, are shown in Figures 6 and 7.

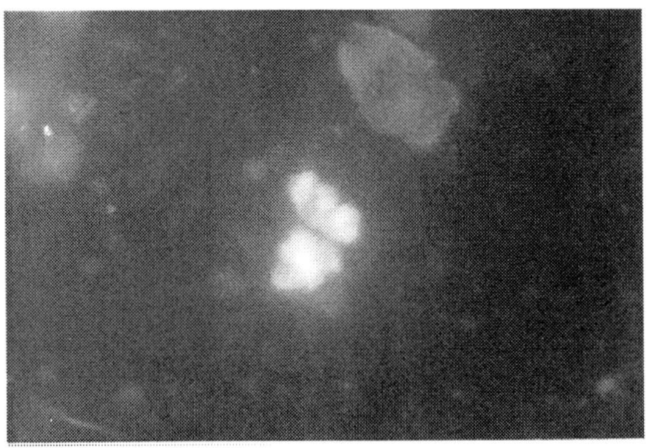

Figure 6. Optical microscopy of fluorescence. Enlargement 1000x (objective: 100, ocular: 10, zoom: 1.25)

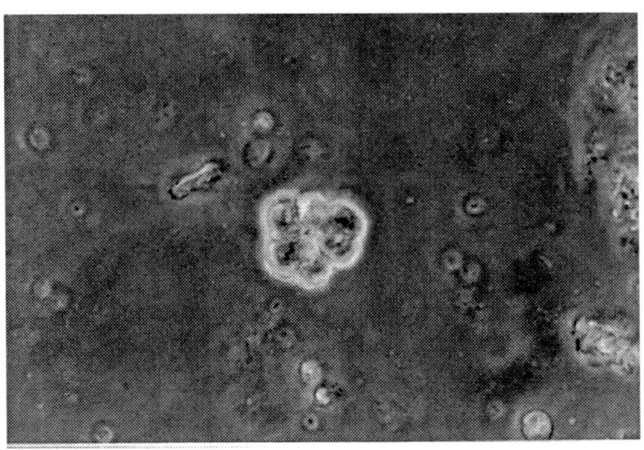

Figure 7. Optical microscopy of phase contrast. Enlargement 1000x (objective: 100, ocular: 10, zoom: 1.25)

Results of common optical fluorescence and phase contrast tests, performed on the material collected from another reactor, on a sample, are shown in Figures 8 and 9.

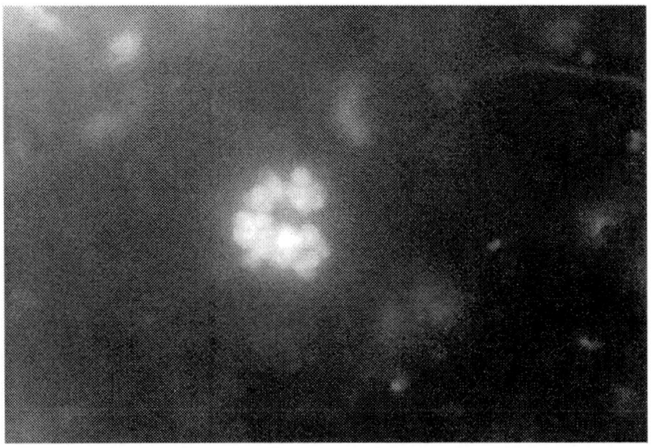

Figure 8. Optical microscopy of fluorescence. Sample: filtering bed of reactor D_3 Enlargement 1000x (objective: 100, ocular: 10, zoom: 1.25)

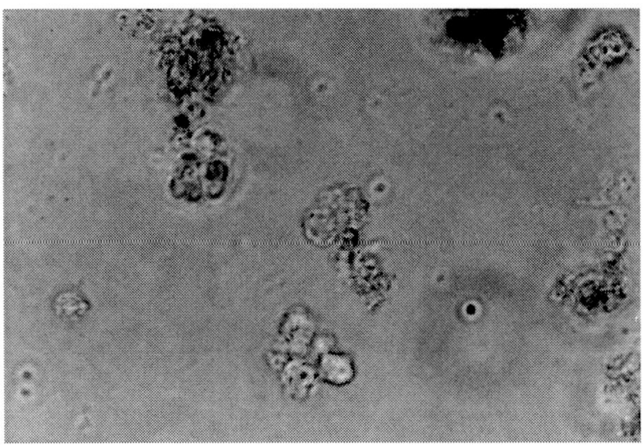

Figure 9. Optical microscopy of phase contrast. Sample: filtering bed of reactor D_3 Enlargement 1000x (objective: 100, ocular: 10, zoom: 1.25)

Results of scanning electron microscopic studies on the samples from filtering bed of reactors are presented in the above figures. The results evidenced the presence of morphology of microorganisms of specie *Methanosarcina sp*, which is further confirmed by the results from common optical microscopy of fluorescence and from phase contrast.

One of the theoretical supports in relation to principal characteristics of specie *Methanosarcina sp* is available in Bergey's Manual of Sistematic Bacteriology (1989), which supports the conclusion drawn by the above study. Referring to the anaerobic treatment process of organic solid waste, the specialized literature indicates predominance of specie of *Methanosarcina sp* kind.

5. Intervening Factors in Anaerobic Treatment Process of Organic Solid Waste

At present anaerobic treatment of solid waste does not present as a well diffused practice in Brazil. Besides the lack of a policy directed to the treatment and final disposal of solid waste produced in the country, there are some operational parameters of the process which require better understanding.

It is known that there are several intervening factors in treatment of organic solid waste, such as: pH, moisture, C/N ratio, temperature, nutrients, toxic substances, amount of inoculum, reactor size, etc.

The reactor size and amount of moisture are the factors of extreme importance since the adjustment of these two factors have direct influence on cost/benefit relation of the process. Lesser the amount of moisture higher will be the amount of organic solid waste contained in a determined volume of reactor, thus decreasing the cost of implantation of the treatment system.

Recently various intervening factors in the process of anaerobic biostabilization of fermentable fraction of organic solid waste have been studied, such as: moisture content, C/N ratio, pH, reactor scale, applied organic load and amount of inoculum.

5.1. Moisture Content

For studying the influence of moisture on anaerobic biostabilization process of organic solid waste 4 anaerobic batch reactors of 20 L capacity each were constructed. The reactors were maintained with 50, 60, 70 and 80% moisture each.

Analyzing the data it was observed that moisture content is of fundamental importance for the anaerobic biostabilization process of organic solid waste. The reactor with 70 wt% moisture developed higher microbiological activity, as it produced 27 L leachate. The reactor containing 60 wt% moisture produced only 3 L leachate and the reactor with 50 wt% moisture during the functioning period of 650 days did not produce any leachate. Therefore, it is verified that in specific case of this work, the moisture percentage in the substrate was of fundamental importance for development of microbial activities and consequently for the biostabilization of the organic material.

Water is a factor of extreme importance in anaerobic digestion, since it is a biological process and all living organisms necessitate water for survival. According to Pelczar *et al.* (1996) water represents about 80 to 90 wt% of a cell, the rest of it is constituted by a combination of other chemical compounds. Water even functions as a fluid medium where most of the cellular biochemical reactions occur. Besides this, water furnishes the substrate and nutrients required by the microorganisms. It is a conducting agent for enzymes and other microbial metabolites important in the process of anaerobic decomposition. The moisture, therefore, contributes decisively for metabolism of the microorganisms (Halvadakis, 1983; Leite, 1997).

A general analysis of performance of the process shows that the efficiency of anaerobic biostabilization of organic solid waste was directly proportional to the moisture percentage. The increase in amount of moisture contributed decisively for the rate of biodegradation of organic material in the reactor.

5.2. C/N Ratio and pH

The influence of C/N ratio and of pH value on the process was evaluated by loading some of the anaerobic reactors with organic solid waste at C/N ratio 20 and other reactors with organic solid waste at pH around 4.0.

According to Tchobanoglous (1993), the ideal ratio of C/N should be about 30. In spite of that, no any problem was verified in relation to excess of nitrogen during the process. The same happened with the pH values, as a good number of reactors during these years were loaded with a substrate at about 4.0 pH. No any corrective measure was applied to these materials and even so no problem related to the low pH value was evidenced. As such, in some of the functioning reactors, even working with substrate presenting acidic pH, an expressive amount of methane in the biogas was verified.

5.3. Reactor Size

Three anaerobic batch reactors of 20, 50 and 200 L capacity each were constructed to evaluate the effect of size during the process. Experimental data show that in general the process efficiency tends to decrease with increase reactor scale. This was verified for most of the monitored parameters, with exception of TVS, whose reduction rate in 20 L reactor was the lowest among the reactors of the experimental system. Analyzing the obtained data it was observed that there exists a strong interaction between reductions of mass of substrate applied in function of reactor size.

The rates of reduction of the principal parameters monitored and the mass applied to the reactors during the realization of the experimental work are presented in Table 5.

Table 5. Rate (%) of reduction of the parameters monitored and the mass applied to the reactors

Reactor	20L	50L	200L
COD	56.8	55.5	51.8
TVS	41.2	52.2	50.3
TKN	67.4	59.0	32.3
Mass of Substrate	30.0	26.0	23.0

From: Leite *et al.*, 2001.

The data presented in Table 5 show that, in general, the efficiency of the process tends to decrease with the increase in reactor size. This was verified for most of the parameters monitored, with exception of TVS, whose reduction rate in reactor A was the least among reactors of the experimental system.

Other workers also have reported similar results in relation reactor size. Borzacconi *et al.* (1997), obtained $1.6x10^{-3}day^{-1}$ hydrolysis constant for a 200 L reactor and $0.8x10^{-3}day^{-1}$ for a 10 m^3 reactor, both treating anaerobically the fermentable organic fraction of municipal solid waste. This shows that during anaerobic biostabilization of municipal solid waste the reactor size influences in performance of the process, seeing that greater the hydrolysis constant shorter will be the time necessary for biostabilization of material in the reactor.

5.4. Amount of Inoculum

In relation to the amount of inoculum used in feeding the reactors four proportions were studied. The amount of inoculum used in preparation of the substrate were 0, 5, 10 e 15wt%. The inoculum used was bovine rumen.

The experimental data obtained showed a strong influence of presence of inoculum on process of anaerobic biostabilization of fermentable organic fraction of municipal solid waste. Among the data obtained, the increase in rate of hydrolysis of particulated material can be verified through the highest quantity of leachate produced by the reactor with 15% of inoculum. Similar trend was shown by increase in biogas production rate with the amount of inoculum. The improvement in performance of the inoculated reactors can be related with the increase in number of native anaerobic microorganisms of rumen which contributed significantly in equilibrating the means of degradation of the organic material present in the reactor. Significant and positive correlations were obtained between the reduction of substrate mass applied to the reactor and percentage of bovine rumen used in preparation of the substrate.

The different mass fractions of substrate applied to the reactors of the experimental system are presented in Table 6.

Table 6. Different mass fractions of the substrate

Inoculum (%)		Mass applied to the reactors (kg)	Mass accumulated in the reactors (kg)	Mass transformed (kg)	Transformation efficiency (%)
1st. phase	0	10.0	8.5	1.5	15.0
	5	10.0	8.0	2.0	20.0
	10	10.0	7.0	3.0	30.0
	15	10.0	6.0	4.0	40.0
2nd. phase	0	25.0	21.7	3.0	13.2
	5	25.0	20.5	4.5	18.0
	10	25.0	18.5	6.5	26.0
	15	25.0	16.0	9.0	36.0

From: Lopes, 2000.

Analyzing the data presented in Table 6, verifies that there is a direct influence of amount of inoculum during the anaerobic biostabilization process. An increase in amount of inoculum used in preparation of the substrate increased the transformation efficiency of the mass applied to the reactors.

5.5. Organic Load Applied

Four reactors were constructed to realization this step, two were monitored with 780kg.m^{-3} organic load and the other two with 900kg.m^{-3}. On basis of the obtained data, it was observed that the reactor with 780kg.m$_{Reator}^{-3}$ organic load and 80% moisture presented better performance. This reactor attained the dynamic equilibrium state in a smaller period, which favors the conversion of substrate in methane, consequently, the biogas produced by this reactor presented higher amount of methane in comparison to the other reactors.

The correlation established between quantities of the waste on dry basis applied to the reactor and rate of leachate production was -0.964. This indicates that these two parameters are strongly related among each other. Working with 900kg.m^{-3} organic load and 70% moisture no sufficient quantity of leachate was liberated for its quantification. The applied organic load presents strong influence on performance of the anaerobic reactors treating organic solid waste with high concentration of solids.

6. Types of Reactors Used in Pilot Scale

One of the fundamental points in relation to anaerobic treatment of organic solid waste is the moisture percentage to be used for the substrate and consequently the reactor configuration. Two alternative technologies in function of moisture content of the substrate for biostabilization of organic solid waste are being studied. The first alternative which still continues in operation is anaerobic treatment of organic solid waste with high concentration of solids. In this case the substrate is being prepared with about 20% of total solids and the biostabilization process is being realized in anaerobic batch reactor. In case of the work with low concentration of solid, the substrate is being prepared with 5% of total solids and the reactor used is being the compartmented anaerobic reactor with non-continuous daily feeding. The objective of the study with these two alternative technologies is to find out that which alternative is more economic in all aspects of the treatment process.

6.1. Anaerobic Batch Reactor: Anaerobic Treatment of Organic Solid Waste with High Concentration of Solids

Anaerobic biostabilization process of organic solid waste was performed in anaerobic batch reactor, with substrate containing about 20wt% of total solids. The substrate was prepared with vegetable solid waste (VSW) collected from free markets and supply centers plus sanitary sewage sludge (SSS).

Chemical characterization data of the vegetable solid waste and sanitary sewage sludge used in preparation of the substrate are presented in Table 7.

The vegetable solid used had slightly acidic character, with an average pH value 4.6 and the moisture content 77.0%. The percentage of VSW was 83.6% and TKN 2.8%, which is slightly higher than the values normally found in municipal solid waste. The sanitary sewage sludge was collected from stabilization pond of a series of anaerobic and facultative ponds that is why the collected sludge presented certain degree of mineralization, with TVS of merely 46.5%.

The experimental system for performing the work was constituted of an anaerobic batch reactor constructed of glass fiber and put on a metallic support, with provision for collection of leachate, recirculation of leachate, collection of biogas for chromatographic analysis, biogas outflow for quantification and the other devices for adjustment of frequently monitored parameters.

Table 7. Chemical parameters of VSW and of SSS used in preparation of substrate

Wastes Parameter	VSW	SSS
pH	4.6	7.5
MA (%)	77.1	80.0
TS (%)	22.9	20.0
TVS (%)	83.6	46.5
COD (%)	43.2	29.8
TKN (%)	2.8	1.3
TP (%)	0.40	-
C/N	16.0	19.8
TOC (%)	46.2	25.8

VSW: Vegetable solid waste; SSS: Sanitary sewage sludge: MA: Moisture amount.

Figure 10 shows prototype of the anaerobic batch reactor used for treating the organic solid waste, of UASB reactor used for treating the leachate resulting from anaerobic batch reactor and the other complementary devices.

The principal operational parameters applied to anaerobic batch reactor are presented in Table 8.

Table 8. Operational parameters

$V_{Reactor}$ (L)	M_{VSW} (kg)	M_{SSS} (kg)	MA (%)	OT (days)	VSW/SSS (%)	M_{COD} (kg)	M_{Total} (kg)
2200	1057.6	264.4	80	270	80/20	114.3	1322.0

VSW: Vegetable solid waste; MA: Moisture amount; OT: Operation time: SSS: Sanitary sewage sludge

Physical and chemical characterization of vegetable solid waste and sanitary sewage sludge was performed and then the substrate was prepared and loaded to the reactor. The total mass substrate applied to the reactor was 1322.0 kg, which corresponds to 1057.60kg vegetable solid waste and 264,40 kg sanitary sewage sludge, in moist condition. As the moisture percentage of the substrate was 80%, on dry basis, it had only 264.4 kg total solids, with a mass of TVS and COD of 221.03kg and 114.2kg respectively. Out of the total mass of substrate loaded in the reactor, it can be observed that 80wt% corresponded to mass of water in its different forms. The presence of moisture in substrate mass favors acceleration of different rates of biostabilization, contributing for reduction of the necessary time so that large part of organic material is bioconverted in final products.

Performance of the biostabilization process was accompanied by quantifying the analytical parameters in liquid and gaseous fractions.

The liquid fraction denominated as leachate, resulting from diverse fractions of moisture present in the substrate and from the natural enzymatic reactions of the process, was collected every fortnight. It was analyzed for the principal parameters of the process such as pH, total alkalinity, volatile fatty acids, total solids and their fractions, TKN and its fractions, COT, BOD_5, total phosphorus and orthophosphate. The chemical and physical methods were used as described by APHA (1995). The gas fraction produced in the process was measured daily and its composition was analyzed by gas chromatography on every alternate day.

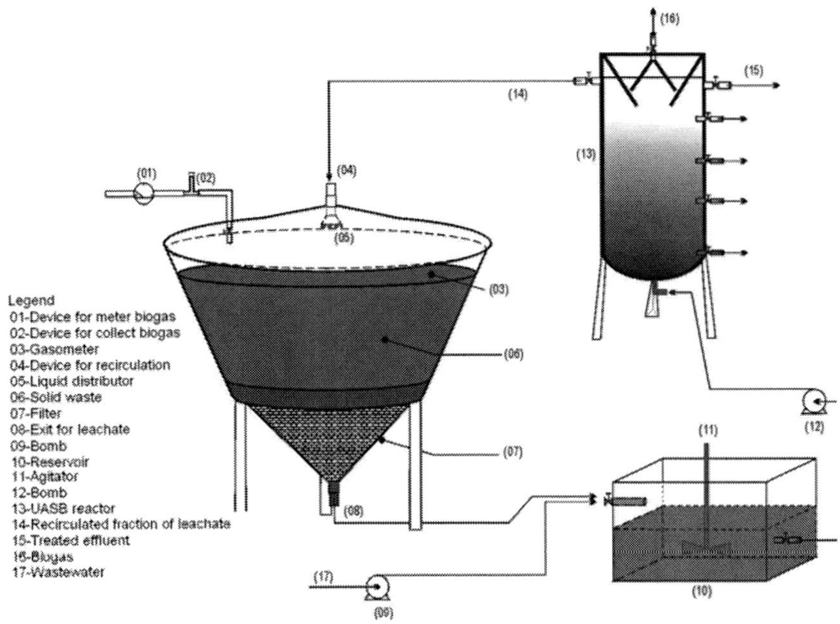

Figure 10. Prototype of the experimental system

The pH of the leachate was monitored for accompanying the buffering or not of the biostabilization process. In the case of substrate composed of chemical compounds of relatively complex chains, the biochemical hydrolysis produces organic acids such as propionate, valerate, isovalerate, butyrate, etc. Therefore, the routine determination of pH expresses the concentration of acidic or alkaline species, contributing in a preliminary evaluation of performance of the process. Figure 11 shows the profile of pH of the leachate during the functioning period of the reactor.

Analyzing the initial stage of the process, corresponding to dissolution of the organic material, the leachate produced contained chemical species with acidic characteristic, maintaining at about pH 5.5 until 170 days of functioning. Only at 200 days of functioning the leachate liquid acquired the neutral stage. The pH value functions as one of the indicators for evaluation of performance of the process.

Leite (1997) when treating fermentable organic fraction of municipal solid waste plus industrial sewage sludge, found that the pH value of the leachated liquids became favorable to the anaerobic biostabilization process only at 180 days of functioning. In that period, the pH value was at about 7.0, with the ratio volatile acids/ total alkalinity of approximately 1.0, demonstrating the presence of equilibrium in different metabolic routes. Borzaconi *et al*

(1997) noted that biostabilization of organic solid waste produced leachate with pH value at about 4.0 during the first 175 days of functioning of the experimental.system.

The volatile acids are resulting from dissolution of particulate material and start behaving as substrate or toxic material for determined species of bacteria responsible for biostabilization of organic material. According to Kroeker (1979) the volatile acids are toxic for the methanogenic bacteria at their concentration of about 2000mgHAC.L^{-1}.

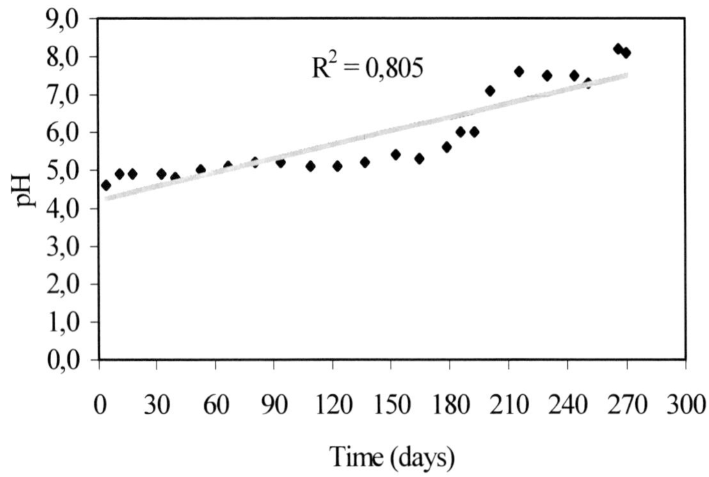

Figure 11. pH profile during the functioning period of the reactor

The total alkalinity is function of the alkalinity due to bicarbonate and volatile acids. In the case of the leachate and in beginning of the biostabilization process, the organic waste produces leachate with low bicarbonate alkalinity and high alkalinity due to volatile acids. As the process enters in the dynamic equilibrium phase, the bicarbonate alkalinity becomes more representative quantitatively. The total alkalinity is determined by equation 4.

$$TA = BA + 0.70\ VA \quad (4)$$

TA: Total Alkalinity (mg$CaCO_3$.L^{-1})
BA: Bicarbonate Alkalinity (mg$CaCO_3$.L^{-1})
VA: Volatile Acids (mgHAC.L^{-1})

Concentration profile of total alkalinity and of volatile acids of the leachate produced during 270 days period of anaerobic biostabilization process of the organic material is shown in Figure 12.

At 33 days of functioning, the leachate presented concentrations of volatile acids and of total alkalinity 18.6 gHAC.L^{-1} and 9.3 g$CaCO_3$.L^{-1}, respectively, establishing a VA/TA ratio as 2.0. It shows a relatively high concentration of volatile acids. These data demonstrate that still there is not a strong symbiotic relation in different species of microorganisms present in the system. After 200 days of functioning, the biostabilization process furnishes buffering agent to the system, since the VA/TA ratio comes down to about 0.7, which shows considerable increase in TA concentration.

For such process, the higher the energetic potential for a certain volume of reactor, better will be the economic viability for the technological alternative developed. Therefore, it is necessary to adjust the operational parameters for increasing the transformation efficiency of the organic material.

COD and TOC were monitored as biostabilization indicating parameters of the organic material. The declining curves of concentration of COD and TOC during the functioning period are presented in Figure 13.

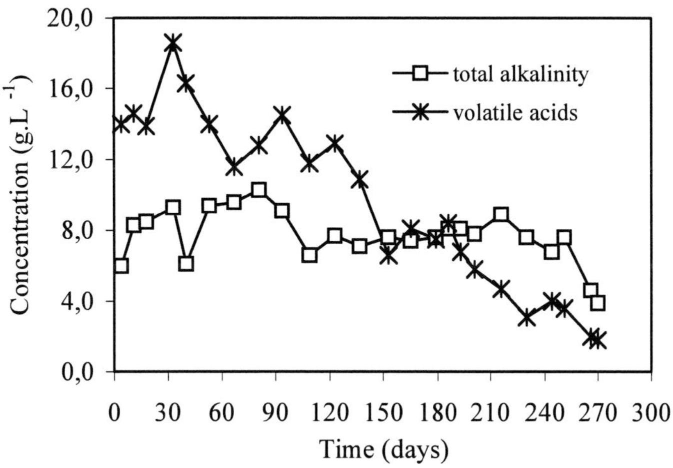

Figure 12. Concentration profile of total alkalinity and of volatile acids

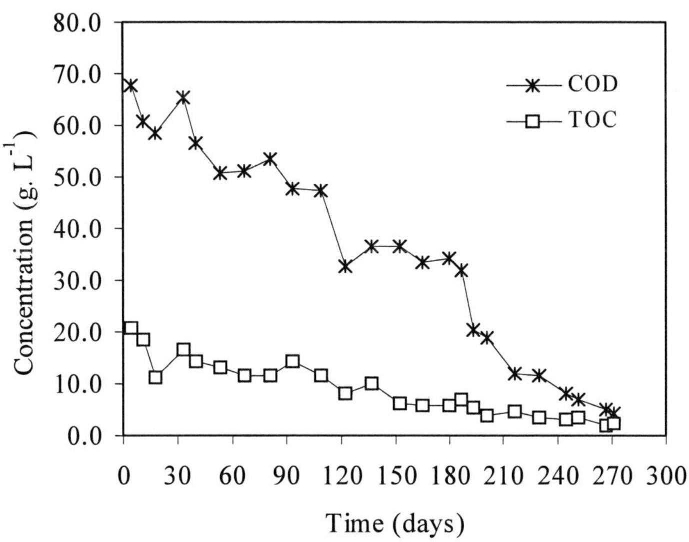

Figure 13. Profile of concentration decrease of COD and TOC

Monitoring the anaerobic batch reactors fed with fermentable organic fraction of municipal solid waste, the amount of affluent COD can be eliminated from the reactor by the leached liquid that, depending on the monitoration system, can lose a considerably significant fraction of the amount of COD, an other fraction is converted in biogas, besides the residual fraction of COD, that is, the amount of COD which remains retained in the partially biostabilized waste. In general, it is expected that out of the total amount of affluent COD, about 70% should be transformed in biogas, in a period of less than 300 days. One of the technological alternatives employed for utilization of the COD eliminated in the leachate, is to recirculate it in the system itself. But for using this alternative, during the initial period of the biostabilization process, a pH correction of acidic leachate will be required. Therefore, initially the leachate should be submitted to a biological treatment process for reducing a part of the organic load for increasing the pH. When its pH attains the value higher than 6, it can be recirculated without any problem.

In this work, quite representative amount of COD was eliminated through the leachate, principally during the initial period of functioning. The decrease in concentration of COD and TOC during the functioning period of the reactor can be seen in Figure 13. The maximum concentration of COD attained the largest peak in the first days of functioning, that is, in the first four days, the average concentration of COD was 65.5 g. L^{-1}. Similar phenomenon happened with the concentration of TOC. The phenomenon occurred in this work may be correlated with strong dissolution of the finely divided solid which should have gone to the lower part of the reactor during its loading and then discharged.

The volume accumulated of biogas produced during 270 days of functioning of the experimental system is presented Figure 14.

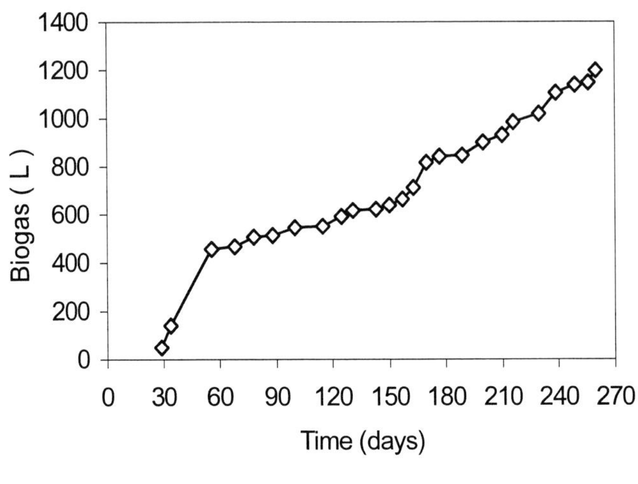

Figure 14. Accumulated production of biogas

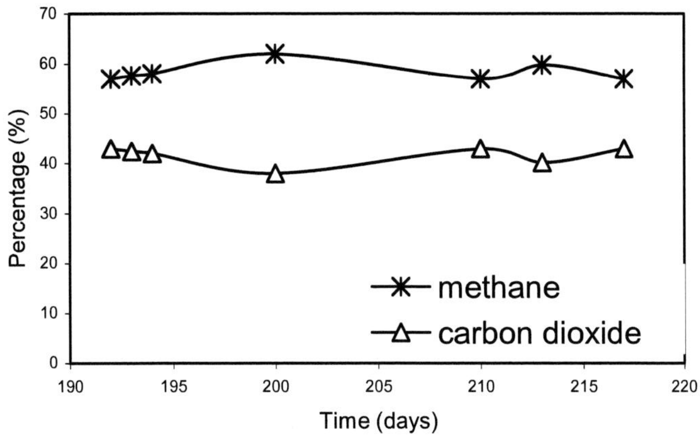

Figure 15. Characterization of the biogas produced

Characterization of the biogas produced in the anaerobic batch reactor is demonstrated in Figure 15. It can be observed from the figure that about 60% of the biogas produced was methane.

It can be pointed out that, for this type of work, the higher the energetic potential for a certain volume of reactor the better will be the economic viability for the technological alternative developed. Therefore, it is necessary that the biostabilization mechanism should be established in such a way that at least 80% of the loaded COD be transformed in a period of about 200 days.

6.2. Compartmented Anaerobic Reactor: Anaerobic Treatment of Organic Solid Waste with Low Concentration Of Solids

Anaerobic digestion process of organic solid waste with low concentration of solids is performed with the substrate containing on an average 5wt% total solids. Problems of technical nature associated with this type of treatment are related with studies of configuration of reactors which has viability for a good efficiency of transformation of carbonaceous material in a relatively short period, the type of substrate that presents good reduction condition of particle size, the dilution liquid for substrate that besides contributing in increase in bacterial density can also furnish carbonaceous material and other types of nutrients and the adjustment of the routine monitoration techniques of the reactor. In general, the anaerobic digestion process of organic solid waste makes possible the daily utilization of the organic solid waste, many times without necessity of intermediate device for storing the waste, which contributes in reduction of operational cost of the system. In relation to the experimental setup, even though working on demonstration scale, a periodic maintenance of the organic waste grinder and of substrate carrying device is necessary for proper functioning of the system.

For treating the organic solid waste with concentration of total solids varying from 4 to 8wt%, a type of reactor is being developed which makes possible a good contact between the waste and the bacterial masses, resulting in about 70% transformation of the affluent total volatile solids in a period varying from 50 to 60 days. The reactor should consist of chambers of equal dimensions and configuration and that each chamber should be responsible in part for a phase of the process of biostabilization of the waste. In addition to this the reactor should be installed with a device for discarding the sludge, principally the sludge with real characteristics of biostabilization, besides an outlet for continuous flow of biogas produced.

Specifically in case of the data presented and analyzed in this section, the experimental system was composed of devices for preparation of organic solid waste, of a feeding tank and of a compartmented anaerobic reactor with 2200 L capacity each. The compartmented anaerobic reactor was constructed of fiber glass and placed on a metallic base. The reactor possessed rectangular geometric configuration and was divided internally by three chambers of equal dimensions. A sampling point was provided in each chamber located at lower and higher levels of the base on a side of the reactor, for accompanying the profile of distribution of the solid material along longitudinal direction of the reactor. Out of the total volume of the reactor, a fraction of about 300 L was destined for storing the biogas produced.

A general scheme of a compartmented anaerobic reactor is shown in Figure 16.

The reactor possessed a total volume of $2.2m^3$, of which $1.85m^3$ was destined for the substrate and the rest of it was for accumulating the biogas. The reactor was monitored in four different phases with organic load applied varying from $11.10kgOSW.m^{-3}.day^{-1}$ to $16.60kgOSW.m^{-3}.day^{-1}$.

The organic solid waste used for feeding the reactor was composed of vegetable solid waste and of sanitary sewage sludge in 80 and 20% proportion respectively. The mixture of these two types of the wastes originated the substrate, which after moisture correction to 95wt% was loaded to the reactor.

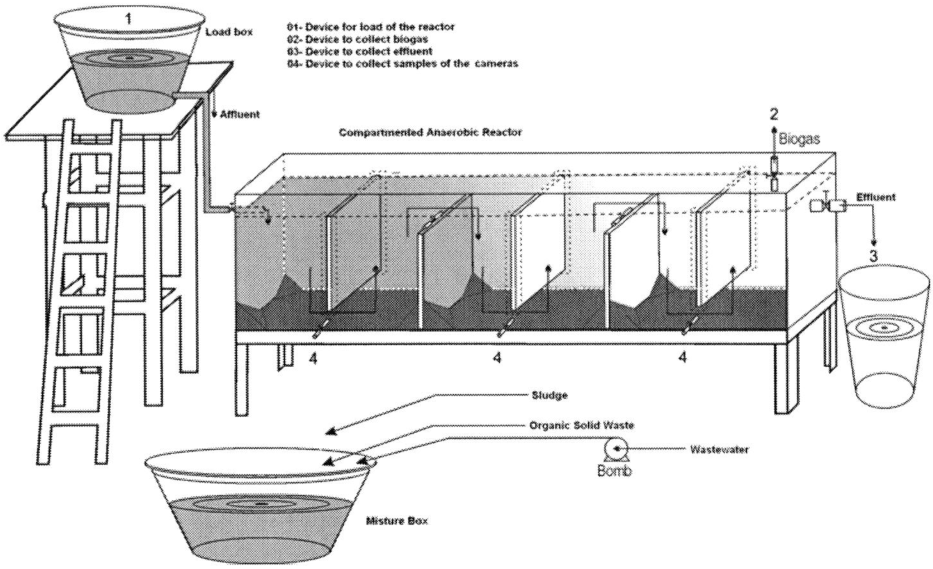

Figure 16. Scheme of a compartmented anaerobic reactor

The monitoration of the experimental system consisted in collection of the organic solid waste, in preparation of the substrate and feeding the reactor daily, besides the systematic accompanying of the waste distribution profile in the internal chambers of the reactor.

The process monitoring was realized in the solid, liquid and gaseous fractions. In case of the solid fraction, the vegetable solid waste, the sanitary sludge and the substrate were always analyzed before feeding the reactor. In case of the liquid fraction the samples were collected weekly and the following parameters were determined: pH, TS and its fractions, total alkalinity, volatile fatty acids, COD, TKN and ammonical nitrogen. For chemical and physical analyses of liquid fractions, the methods given in APHA (1995), with due methodological adaptations, were used. In gaseous fraction, the quantification of the biogas was realized continuously, where as the qualitative characterization was done by gas chromatography two times every week.

One of the parameters which is being investigated with sufficient interest in anaerobic biostabilization process of organic solid waste is the moisture content. The excellent adjustment of moisture content for performance of the process will contribute technically for attaining a higher index of transformation of the organic material and will reduce the cost of implantation of the experimental system, thus equilibrating favorably the cost/benefit ratio of the treatment. The present work using 95% of moisture content was therefore realized.

The principal results obtained from functioning of the compartmented anaerobic reactor containing organic solid waste with 5% total solids are presented and discussed here for the purpose of delineating the project parameters for applying on this type of project at real scale.

The principal operational parameters applied on compartmented anaerobic reactor are presented Table 9.

Table 9. Operational parameters applied to the reactor

V_R (m^3)	M_{VSW} (kg)	M_{SSS} (kg)	M_T (kg)	$R_{RSV/LES}$ (%)	OP (days)	MC (%)	M_{COD} (kg)	M_{TVS} (kg)
2.2	1480.0	370.0	1850.0	80.0/20.0	90.0	95.0	35.6	70.3

OP: operation time; MC: moisture content; V_R: volume of reactor

1,480kg of typically vegetable solid waste plus 370kg of sanitary sewage sludge, totaling to 1,850kg were loaded in the reactor. Considering the presence of 95% moisture, out of the total mass of the substrate only 92.5kg corresponded to total solid mass. From the mass of solid material on dry basis 70.3kg was total volatile solid and 35.6 kg was COD.

Physical and chemical characterization data of the substrate loaded in the reactor are presented in Table 10.

Table 10. Physical and chemical parameters of the substrate

Parameter	pH	MC (%)	TVS (%)	COD (%)	TKN (%)	TOC (%)	C/N	TA ($gCaCO_3L^{-1}$)	VFA ($gH_{AC}.L^{-1}$)
Value	5.6	95.0	76.0	38.5	3.5	42.2	12.1	6.6	7.4

Profile of pH of affluent and effluent of the reactor during 140 days of functioning period with the applied organic load of 11,10kgOSW.m^{-3}.day^{-1} is shown in Figure 17.

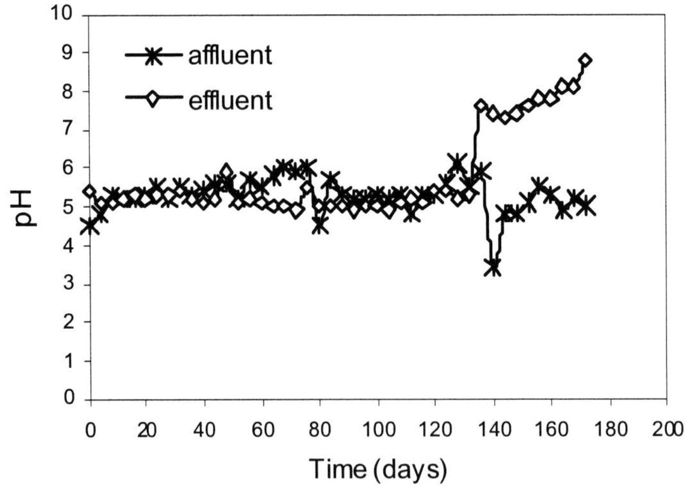

Figure 17. Profile of pH of affluent and effluent material of reactor

Analyzing the pH profile of the affluent material of the reactor, it is observed that the pH values vary from 3.4 to 6.0 and this variation of pH is associated with heterogeneity of the fermentable organic fraction of the organic solid waste. In the effluent material, until 140 days of functioning the pH values were always around 5.0, increasing after the 140 days of functioning for the values higher than 7.0, reaching up to the level of 8.0.

The profiles of total volatile solids of the affluent substrate and of the effluent material from the reactor are demonstrated in Figure 18.

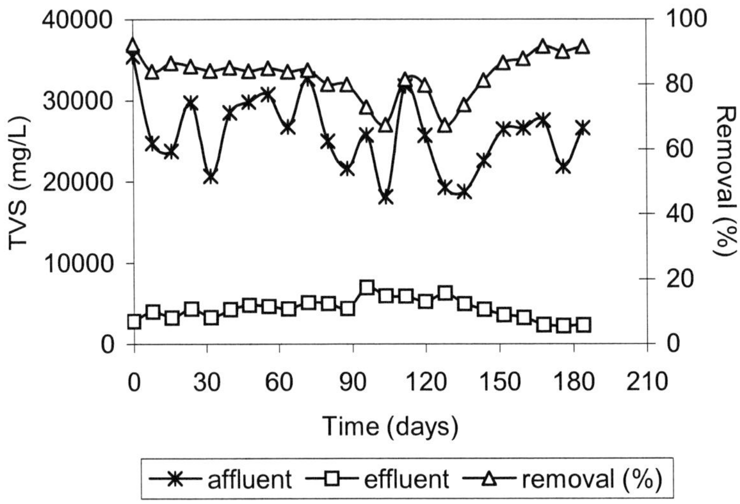

Figure 18. Profile of TVS in affluent material and effluent from reactor

Analyzing the profile of TVS in the affluent material and effluent from reactor presented in Figure 18, it was observed that the concentration of TVS in affluent material during the functioning period varied from 18134mg.L^{-1} to 31750mg.L^{-1}, whereas in the effluent material the concentration of TVS varied from 2190 to 6972mg.L^{-1}, attaining a removal efficiency of about 80% of TVS.

Profile of chemical oxygen demand of the affluent substrate and of the effluent from the reactor is presented in Figure 19.

Figure 19 shows that concentration of COD in affluent material of the reactor varied from 9713.0 to 32744.0 mg.L^{-1}. The variation of COD concentration of the substrate is many times associated with heterogeneity of the substrate, besides this difficulty in sampling. In the effluent material from the reactor, the concentration of COD varied from 1131.0 to 10558.0 mg.L^{-1}, attaining an average transformation efficiency of 60%.

Biogas and methane production rates in function of mass of COD applied are presented in Figure 20.

The production rate of biogas observed was directly proportional to the applied organic load. Some of the points obtained showed production rate sufficiently higher than the average value, which can be ascribed to fluctuations in atmospheric temperature, that is, with an accentuated increase in temperature there was increase in biogas production rate. The rates of production of methane were directly proportional to the rates of production of biogas. The volume of biogas produced contained 60% of methane in it. Taking the last three phases of this work as reference, the energetic percentage obtained was approximately 0.25Nm^{3}CH$_4$.kg^{-1} of COD applied or 37NLCH$_4$.hab^{-1}.day^{-1}, which represents a sufficiently significant contribution in national energetic matrix.

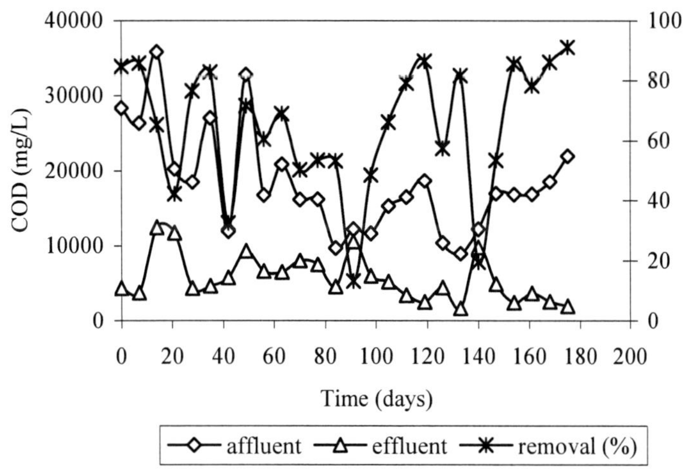

Figure 19. Profile of COD of affluent material and effluent from reactor

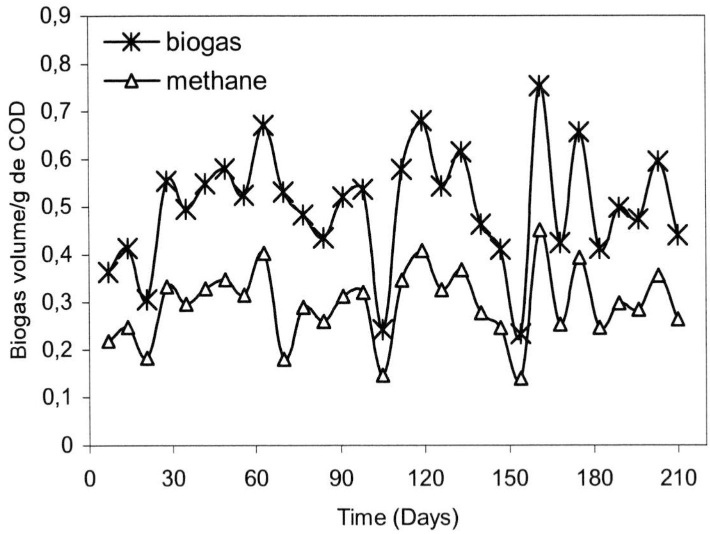

Figure 20. Biogas and methane production rates

In reactors of this configuration, treating organic solid waste with low concentration of solids, the principal factors which influenced the transformation of the carbonaceous material are the retention time of the solids, the physical and chemical composition of the substrate and the rate of organic loading. Therefore, in work of this nature whose principal objective is to achieve an excellent transformation of carbonaceous material in a relatively short period of time, it is necessary to investigate the optimum concentration of nutrient material in the waste to be treated, the granulometry of solid material present in the substrate and the other intrinsic factors of the anaerobic biostabilization process.

In general, the anaerobic treatment process of organic solid waste with low concentration of solids in compartmented reactor, did not present operational problems when applied for organic load up to $0.25kgCOD.m^{-3}.day^{-1}$. The process retained about 90% of the carbonaceous material and produced an energy potential of $0.25Nm^{3}CH_{4}.kg^{-1}COD$.

7. CONCLUSION

The anaerobic treatment process can be applied to destination of organic solid waste, being necessary only to adjust the amount of moisture in the waste and in some cases the C/N ratio. As for adjustment of moisture amount, one can work on substrates with total solid concentrations varying from 5 to 20%. What requires more study is the configuration of the reactor for treating these substrates and, consequently, the adjustment of the cost/benefit ratio of the treatment process, taking into consideration the questions of environmental approach.

The principal point to be considered when selecting the anaerobic treatment process for organic solid waste is related to separation of this fraction from other fractions since the selective collection system is not in practice, or there should be some other mechanism for complete separation of the organic solid from the other types of waste.

When the use of an anaerobic treatment process is proposed, the principal objective should be the utilization of methane gas as an energy source for employing domestic, agricultural or industrial activities, thus contributing to the principal energy matrix. Besides this contribution, the partially biostabilized waste can still be used in soil for plantation after passing through the vermicompost treatment process.

In general, it can be concluded that the various intervening factors existing in the organic solid waste treatment process in anaerobic reactors have made difficult the application of this type of process at real scale. Thus, one should be very careful while projecting this type of treatment system.

REFERENCES

APHA, Standard Methods for Examination of Water and Wastewater. 19th edition, American Public Health Association, American Water Works Association, Washington, DC, 1995.

Barlaz, M. A; Ham, R.K. *Jour. Envir. Engineering*, 1989, 115, 1088 – 1101.

Bergey's manual of systematic bacteriology. 1st edition, Williams and Wilkins, Baltimore, 1989.

Bitton, G. *Wastewater Microbiology*, 2nd edition, John Wiley and Sons, Inc, New York, 1999.

Borzacconi, L.; López, I.; Anido, C. *Wat. Sci. Technology*, 1997, 36, 479-484.

Bouallagui, H.; Touhami, Y.; Chaikh, R.B.; Hamdi, M. *Pros. Biochemistry*, 2005, 40, 989-995.

Bryant, M.P. *Jour. Ani. Science*, 1979, 48, 193-201.

Chernicharo, C.A.L. *Reatores anaeróbios*; Vol. 5, DESA/UFMG: Belo Horizonte/MG, 1997.

Christ, O.; Wilderer, P.A.; Angerhöfer, R.; Faulstich, M. *Wat. Sci. Technology*, 2000, 41, 61-65.

Gujer, W.; Zehnder, A.J.B. *Wat. Sci. Technology*, 1983, 15, 127-167.

Halvadakis, C.P. *Methanogenesis in solid waste landfill bioreactors*. Stanford University, 1983. Dissertation.

Kroeker, E. *J. Jour. WPCF*, 1979, 51, 718-728.

Lay, J. J.; Li, Y. Y.; Noike, T. *Wat. Sci. Technology*, 1998, 38, 177-184.

Leite, V.D. Processo de tratamento anaeróbio de resíduos sólidos urbanos inoculados com lodo industrial. São Carlos, *EESC/USP*, 1997. Tese de Doutorado.

Leite, V. D.; Lopes, W. S.; Prasad, S. *Rev. Bras. Eng. Agríc. e Ambiental*, 2001, 5, 119-123.

Leveau, J. Y.; Bouix, M. In: *Biotecnologia*; SCRIBAN, R. Editora Monole LTDA, São Paulo, 1985, 103 – 123.

Lopes, W. S. Biodigestão anaeróbia de resíduos sólidos urbanos inoculados com rúmen bovino. Campina Grande-PB. *PRODEMA/UFPB/UEPB*. Dissertação de Mestrado, 2000.

Lopes, W.S.; Leite, V.D.; Prasad, S. *Jour. Sol. Was. Tech. Management*, 2003, 29, 108-117.

Lopes, W.S.; Leite, V.D.; Prasad, *S. Bior. Technology*, 2004, 94, 261-266.

Metcalf and Eddy. *Wastewater engineering treatment and reuse*; 4th Edition; McGraw-Hill: New York, 2003.

Nascimento, R. A. Desempenho do reator anaeróbio de manta de lodo utilizando efluentes líquidos da indústria alimentícia. Campinas, *DHS/UNICAMP*. Dissertação de Mestrado, 1996.

Oremland, R.S. In: *Biology of Anaerobic Microorganism*; ZEHNDER, A.J.B.; John Wiley and Sons. Wageningen, 1988.

Pavlostathis,S.G. *Jour. Envir. Engineering*, 1998, 114, 125-132.

Pelczar, M.J. et. al. *Microbiologia: conceitos e aplicações*. Editora Makron Books, São Paulo/SP, 1996.

Peres, C. S.; Sanches, C. R.; Matumoto, C.; Schmidell, W. Anaerobic Biodegradability of the Organic Components of Municipal Solid Waste (OFMSW). In *Proceedings of the VI International Symposium of Anaerobic Digestion*. São Paulo, Brazil, 1991, 12-16.

Sanders, M.; Geerink, M.; Zeemam, G.; Lettinga, G. Wat. Sci. Technology, 2000, 41, 17-24.

Santos, *J.M. Microscopia Eletronica de Varredura Aplicada as Ciências Biológicas.* FCAV/UNESP – Jaboticabal/SP, 1992.

Schmidell Netto, W. Cinéticas dos processos fermentativos. IV Curso de tratamento biológico de resíduos, *FURB/UFSC*. Florianópolis, 1999.

Stronach, S. M. et al. *Anaerobic digestion Process in Industrial Wastewater Treatment.* Public Helath Engineering Laboratory. London, 1986.

Tchobanoglous, G. et al. *Solid Wastes: Engineering principle and management Issues.* McGraw-Hill Book Company, Ney York, 1991

Van Haandel, A.; Lettinga, G. *Tratamento anaeróbio de esgotos - um manual para regiões de clima quente*. EDUFPB: Campina Grande/Brazil, 1994.

Van Lier, J. B.; Tilche, A.; Ahring, B. K.; Macarie, H.; Moletta, R.; Dohanyos, M.; Hulshoff Pol, L. W.; Lens, P.; Verstraete, W. *Wat. Sci. Technology*, 2001, 43, 1-18.

Vazoller, R. F. Microbiologia e Saneamento Ambiental. IV Curso de tratamento biológico de resíduos, *FURB/UFSC*. Florianópolis, 1999.

Viñas, M. Critérios de diseño y escalado de reactores anaerobios. In: *Taller y Seminario Latinoamericano Tratamento Anaerobio de Aguas Residuales*. Montevideo, Uruguay, 1994, 111-123.

Zehnder, A.J.B.; Stumm, W. In: *Biology on Anaerobic Micoorganisms*; Zehnder, A.J.B.; Agricultural University Wageningen: Netherlands, 1988.

In: Solid Waste: New Research
Editors: A. S. Martin and J. D. Suarez
SBN: 978-1-60456-809-7

Chapter 8

THE CHEMICAL PROPERTIES OF MUNICIPAL SOLID WASTE INCINERATOR ASHES AND THE EFFECTS OF THEIR UTILIZATION AS LANDFILL COVER ON LANDFILL BIOSTABILIZATION

***Huang-Mu Lo**[*a]**, Min-Hsin Liu**[a]**, Chao-Yang Lin**[b]**, Wen-Fung Liu**[c]**, Tzu-Yi Pai**[a]**, Chun-Hsiung Hung**[b]**, Pin-Hung Cheng**[d]**, Yuan-Lung Liao**[a]**, Tsu-Ying Fu**[a]** and Chao-Chan Yang**[a]*

[a]Chaoyang University of Technology, Taiwan, R. O. C.
[b]National Chung Hsing University, Taiwan, R. O. C.
[c]Feng Chia University, Taiwan, R. O. C.
[d]Taipei University of Science and Technology, Taiwan, R. O. C.

ABSTRACT

This article investigated the properties of municipal solid waste incinerator (MSWI) ashes and the effects of their addition on the municipal solid waste (MSW) anaerobic digestion as co-disposed or co-digested with MSW in landfill or digester. Five anaerobic bioreactors with the size of 1.2 m height and 0.2 m diameter were employed to conduct the experiment. Four layers were arranged each with 6.5 liter of MSW and anaerobic seeded sludge mixture covered by 2.5 liter of MSW and anaerobic seeded sludge mixture blended with the designed ashes added ratios as well as the control bioreactors without ashes addition. The synthetic MSW used in this experiment was typical of organic fraction of MSW and was comprised of newspaper, food waste, office paper and hay etc. MSWI ashes were obtained from a mass burning incinerator in central Taiwan. Also, the seeded anaerobic sludge was taken from a municipal wastewater treatment in central Taiwan. As the experiment was proceeded with, the leachate of 100 mL was recirculated per day and another more 100 mL one was collected and filtered for parameters analysis such as pH, conductivity, alkalinity and chemical oxygen demand (COD) etc. In addition,

* Corresponding author. Tel.: +886-4-23323000 ext 4469; fax: +886-4-23742365. E-mail address: hmlo@cyut.edu.tw (Huang-Mu Lo)

the gas production rate was recorded every day to measure the bacterial activity in the MSW biodegradation. From the results, it showed that 10 and 20 g l^{-1} fly ash added (g ashes addition per liter MSW ratios) bioreactors and 100 g l^{-1} bottom ash added bioreactors were found to enhance the gas production rate and the soluble concentration of alkali metals such as Ca, Mg, K, and Na as compared to the control one. The six soluble heavy metals of Cd, Cr, Cu, Pb, Ni and Zn in leachate were also found to be under inhibitory concentration for anaerobic digestion. Other trace metals such as Co and Mo etc were assumed to serve as the stimulatory micronutrients rather than to exert inhibitory effects on the microorganisms in the MSW anaerobic digestion.

Keywords: anaerobic digestion, landfill, MSW, MSWI ashes

INTRODUCTION

Municipal solid waste (MSW) incinerator (MSWI) has been practiced in the MSW treatment as important as landfill, composting and resource recovery and recycling. This is because that the likely landfill site is getting more difficult to find for the MSW disposal particularly occurred in the dense population and lesser appropriate land for landfill in Taiwan. MSWI can take the advantage of MSW weight or volume reduction and the recovery of steam and electricity. However, the residues such as bottom ash and fly ash generated were still the primary environmental concern. Thus, they need to be carefully treated to prevent the potential secondary pollution. Particularly, the released metals, ions and other compounds if not properly treated might be toxic to the human health and ecological environment when utilized in different purposes. Therefore, MSWI residues treatment and disposal has become another environmental task in MSW treatment. MSWI residues have been utilized as aggregate, backfill, soil amendment and solidified for permanent landfill storage as well (Bertolini et al., 2004; Deschamps, 1998; Hjelmar, 1996). In addition, MSWI bottom ash has been practiced as landfill cover, however, the baseline information and landfill mechanisms were not fully clear. Only few research investigated the bottom ash co-digested with MSW in the semi-batch anaerobic bioreactor (Lo, 2005). This study showed that bottom ash addition could provide the alkalinity suitable for the anaerobic digestion thus enhancing the gas production rate. Another investigation also presented that $NaHCO_3$ could provide alkalinity and facilitate the MSW anaerobic digestion process (Ağdağ and Sponza, 2005).

As regard to MSWI fly ash, it is thought that using fly ash as landfill cover is an aggressive challenge for the landfill practice. It is further noted that fly ash has the potential to release the heavy metals which is considered to be hazardous to the soil and ground water if disposed improperly or leaked from the faulty geo-membrane liner in the bottom of landfill. The soil and ground water contamination by the release of MSWI ashes in landfill might have the potential to cause detrimental effects on human health and ecological environment. Therefore, using fly ash as landfill cover might be an aggressive challenge needing further theoretical and experimental investigation. In addition, it is noted that MSW landfill needs a long term period to obtain the MSW biostabilization. Thus, anaerobic landfill bioreactors were employed to conduct the experiment of MSW and MSWI ashes co-disposal to obtain the results in a short term. Particularly, the pH, alkalinity, COD, metals and gas production in the bioreactors were measured to assess the MSW biostabilization in landfill bioreactor simulating the real landfill site.

MATERIALS AND METHODS

MSWI Ashes

The MSWI bottom ash and fly ash required for the bioreactor experiment were derived from a mass burning MSWI in central Taiwan (Figure 1). Generally, bottom ash was obtained from the incinerator stoker which is the major part for the MSW incineration as well as the siftings leaked from the stoker. Fly ash was derived from the air pollution control device (APCD) such as semi-scrubber and bag filter. Fly ash has been measured to have smaller particle size as compared to that of the bottom ash by sieve analysis. Thus, it is expected that fly ash particle contains higher specific surface than that of bottom ash leading in the potential higher release of metals, ions and compounds. These released metals, ions and compounds might further influence the different utilization purposes. Particularly, the released heavy metals were reported to have the potential to cause the human health and ecological environment. The metal total content in MSWI ashes was characterized as Figure 2.

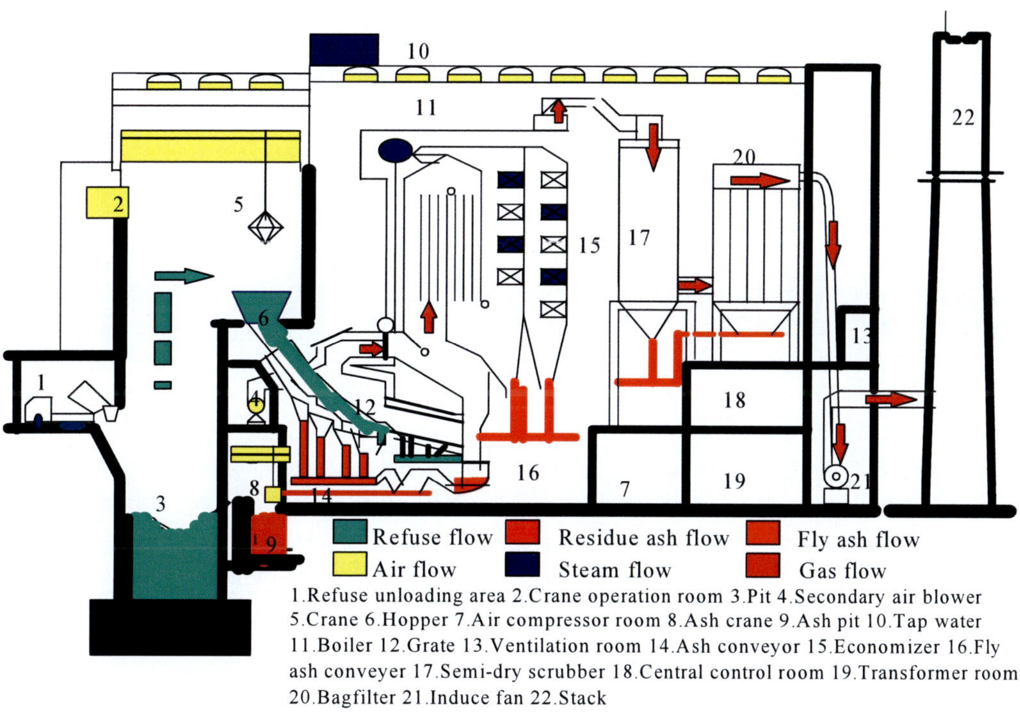

Figure 1. The treatment process of mass burning MSW incinerator

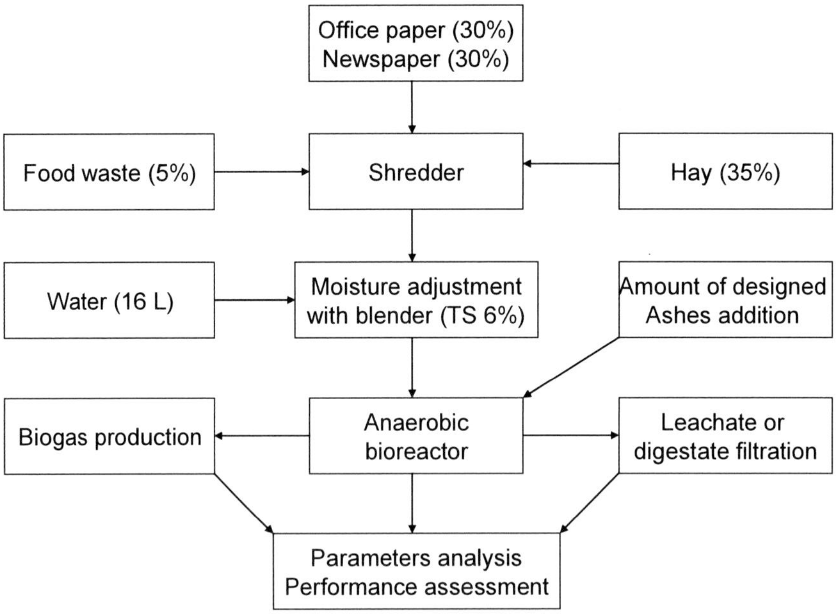

Figure 2. Schematic diagram showing the preparation of feedstock, its component composition, and bioreactor outputs and analysis

MSW

In order to reduce the interference of potential hazardous materials in MSW, synthetic MSW was prepared artificially in the laboratory. The MSW was comprised of newspaper (30%), office paper (35%), food waste (5%) as potato and hay (30%) etc. These constituents were typical of major organic fraction of MSW. With the measurement by elemental analyzer, the chemical composition consisting of C, H, O, N and others were about 46 %, 6 %, 41 %, 1.4 % and 5.6 % respectively. These constituents contained typical organic fraction of MSW and were easily to be biodegraded by microorganisms. For the convenience of microbial attack in the decomposition process, synthetic MSW was shredded by a shredder to pieces of less than 5 mm. Then, it was blended with distilled water to make a total solid (TS) of 6%. Although this percent of TS was lower than that of real landfill MSW, the carbon content is near the same. In addition, this investigation focused on the effects of MSWI ashes co-disposed with MSW to examine the metals and ions release on the MSW decomposition and their possibility of utilization as landfill cover. The MSW preparation procedure and metal total content in MSW were listed as Figure 2 and Figure 3.

Anaerobic Seeded Sludge

To ensure the MSW biodegradation occurred without retardation, anaerobic seeded sludge was taken from a municipal waste water plant in central Taiwan. This plant collects the municipal waste water from Taichng city. The anaerobic sludge showed efficient bacterial activity that was beneficial for the startup and biodegradation of MSW in the anaerobic bioreactor. The metal content of anaerobic seeded sludge was indicated in Figure 3.

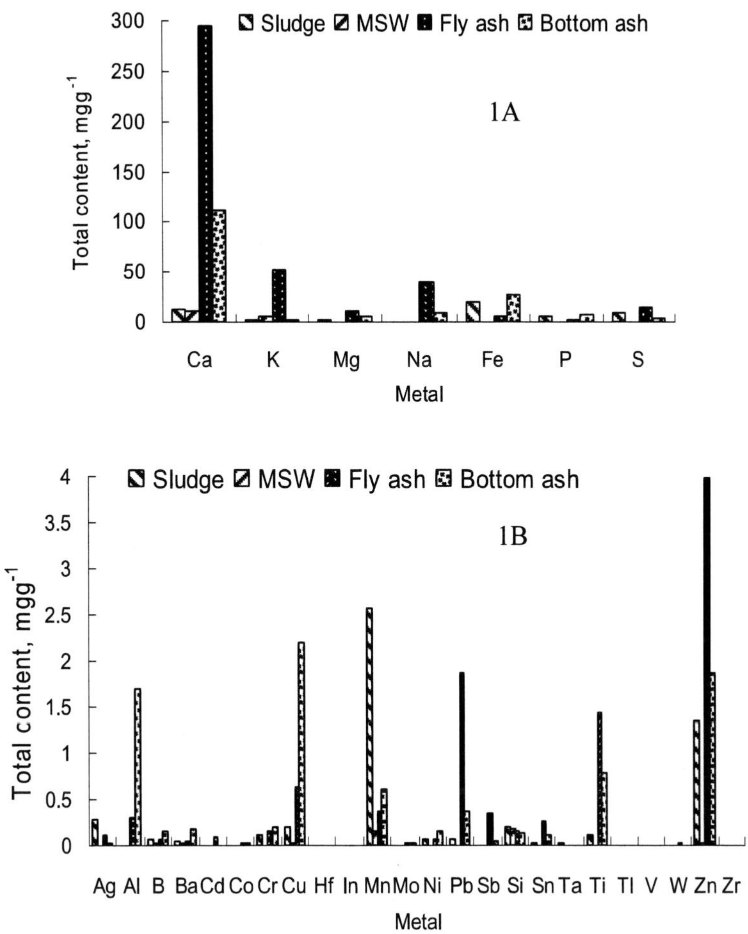

Figure 3. The selected metal content in sludge, MSW, fly ash and bottom ash

Anaerobic Bioreactor

In order to obtain the effects of MSWI ashes co-disposed with MSW in a short term, five anaerobic bioreactors (Figure 4) with the size of 1.2 m height and 0.2 m diameter were employed to conduct the experiment. Four layers of co-disposal simulating the landfill site operation were arranged. Two of them were set up for bottom ash co-disposal with MSW and the other two were for fly ash co-disposal with MSW. The fifth one was used to compare the results without any ashes addition. The designed added ratios of bottom ash were 100 and 200 g l^{-1} (100 and 200 g addition per liter MSW) and of fly ash were 10 and 20 g l^{-1} respectively. In these bioreactors, it consisted of 22 liter MSW and 12 liter of anaerobic sludge accompanying by the designed ashes addition. Each layer within the four ones of anaerobic bioreactor contained 6.5 liter of MSW and seeded sludge mixture and was covered with 2 liter of MSW and seeded sludge mixture blended with the designed MSWI ashes addition as previously stated. The 100 ml leachate at the bottom of the bioreactor was recirculated each day. In addition, another more 100 ml leachate was collected and filtered for parameters analysis such as pH, alkalinity, chemical oxygen demand (COD), total solid (TS), volatile solid (VS) and metals daily or weekly. Gas production was also measured with gas collector daily. The measured parameters and gas production rate can be used to assess the performance of the anaerobic bioreactors.

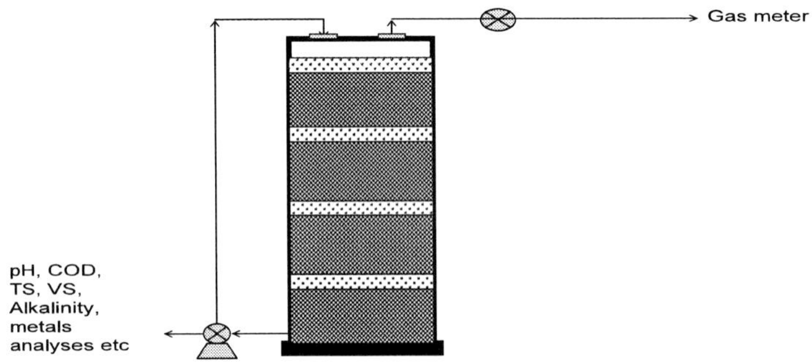

Figure 4. The schematic diagram of anaerobic bioreactor in the experiment

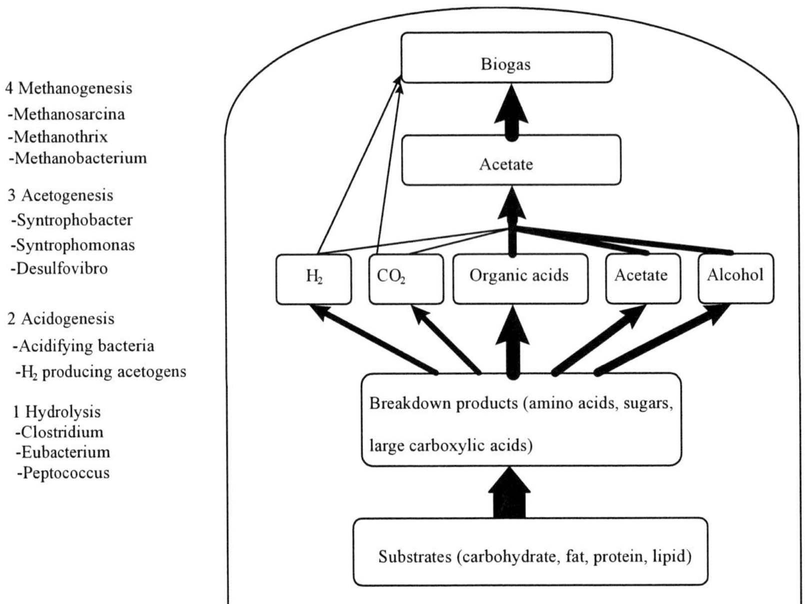

Figure 5. Biochemical process in anaerobic digestion (Dichtl, 1997; House et al., 1997; Bhatti et al., 1996)

pH was measured by a pH meter. Alkalinity and COD were measured with titration methods. TS and VS were analyzed with the oven at 105 and 600°C respectively. All these analysis methods followed the standard methods for the examination of water and wastewater (AWWA, 1995). Microorganisms play an important role on MSW anaerobic digestion process (Figure 5). Molecular biotechnology of FISH was used to characterize the selected methanogens such as *Methanobacteriales*, *Methanococcales*, *Methanosarcinales* and *Methanomicrobiales* and the target probes were MB310, MC1109, MSMX860 and MG1200 (Raskin et al., 1994) respectively. Methanogens detection could provide an indication of methanogenesis extent of MSW anaerobic digestion as the MSWI ashes were added.

RESULTS AND DISCUSSION

Parameters Analysis

From the experimental results, it showed that pH trends of bioreactors in leachate (Figure 6A) have shown lower values at the beginning in the five bioreactors. Thereafter, they reached to the steady values of about pH 6.2-7.5. The pHs in the bottom ash added bioreactors were found higher than that in fly ash added and control ones specifically found in the 200 g l^{-1} bottom ash added bioreactors with the pHs higher than 7.5 after about month six. Further, the lower pHs below 6 found in the five bioreactors at the beginning except first week was thought that large amount of VA was produced in the first stage of MSW hydrolysis leading to the higher volatile acids and lower pHs. However, the pHs returned to about pH 6.2-7.5 soon possibly due to the alkali metal hydroxides and carbonates release which could provide the buffer alkalinity and neutralize VA as a function of pH. This range of near neutral pHs were thought to be suitable for the anaerobic digestion (Lo, 2005; Parkin and Owen, 1986).

Conductivity was found higher in the ash added bioreactors than the control one (Figure 6B). Conductivity in the 20 and 10 g l^{-1} fly ash added bioreactor was found to be about three times and two times that of control one respectively. This is because that higher fly ash addition could release higher different kinds of ions such as metals and chlorides etc resulting to a higher conductivity. As regarding to alkalinity, the alkalinity concentrations in the five bioreactors had the same trends between 500 and 3500 mg l^{-1} and were found to be in the suitable range for the anaerobic digestion as described by Parkin and Owen (1986) and can be seen in Figure 6C.

VA in the five bioreactors was found to decrease except that found to increase in the 20 g l^{-1} ash added bioreactor between week two and week five (Figure 6D). From the third month on, the VA started to increase slightly. VA found higher in the first month was thought that the MSW was hydrolyzed in the first digestion stage resulting to a large amount of VA production such as acetate, propionate, butyrate, valerate etc. VA concentrations were then maintained in a steady state within the five bioreactors from month two to about month three. In the mean time, VA was converted to the methane and carbon dioxide by methanogenic bacteria as the gas production rate was found higher in this period. In addition, the VA in the MSWI ashes added bioreactors was found slightly higher than that in the control bioreactors at the beginning. From the fourth month on, the VA in the five bioreactors was found to increase slightly. This phenomenon indicated that the methanogenic activity began to decrease possibly due to the MSW anaerobic digestion completion and the lesser MSW substrate left to be utilized.

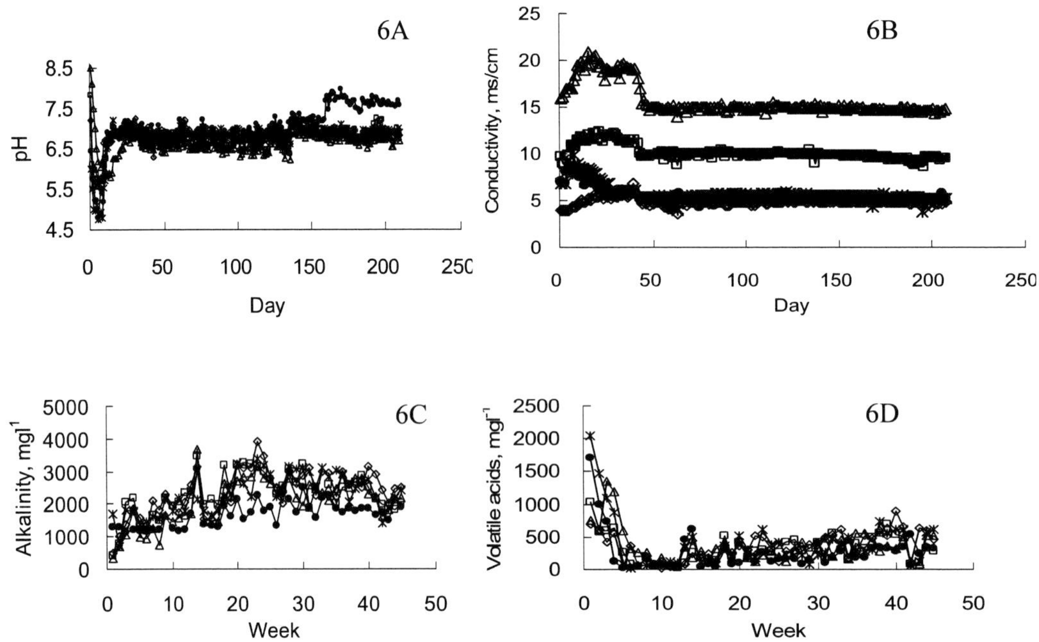

Figure 6. The trend of pH, conductivity, alkalinity and volatile acids in leachate of MSW anaerobic digestion by five different addition ratios of MSWI fly ash added, bottom ash added and the control bioreactors (◇: control bioreactor without ash addition; □: 10 g l-1 fly ash added bioreactor; △: 20 g l-1 fly ash added bioreactor; ＊: 100 g l-1 bottom ash added bioreactor; •: 200 g l-1 bottom ash added bioreactor).

GAS PRODUCTION RATE AND METALS RELEASE

Released metals have been found to influence the MSW anaerobic digestion. The gas production rate in the MSWI ashes added bioreactors was found higher as compared to that in the control one (Figure 7). Higher gas production rate in the ash added bioreactors was thought that the release of alkali metals such as Ca, Mg, K and Na and the associated OH^{-1} and CO_3^{-2} have provided the pH buffer in neutral band which facilitated the anaerobic digestion. In addition, six heavy metals of Cd (0-0.001 mgl^{-1}), Cr (0-0.014 mgl^{-1}), Cu (0-0.05 mgl^{-1}), Pb (0-1.75 mgl^{-1}), Ni (0-0.025 mgl^{-1}) and Zn (0-0.05 mgl^{-1}) in leachate were measured to be under inhibitory soluble concentration. It is further noted that trace metals such as Co, W, Ni etc might serve as the micro nutrients (Takashima and Speece, 1990; Kayhanian and Rich,1995; Paulo et al., 2004; Fish, 1999; Zandvoort et al., 2004). Thus, the gas production rate was facilitated and the MSW digestion was accelerated. Moreover, microorganisms of selected methanogens were found higher in the period of higher gas production rate within the ashes added bioreactors. Microorganisms seemed to be stimulated by the suitable metals release which was beneficial to the anaerobic digestion. Therefore, MSW anaerobic digestion efficiency was enhanced and the gas production rate and MSW biostabilisation were increased. It was further noted that the highest gas production rate in all bioreactors seemed to take place about on day 30. The gas production rate of the ashes added bioreactors on the peak was found to be about 9000 ml per day nearly two times that of 4500 ml per day found in the control one. The gas production rate in the bottom ash added bioreactor decreased quickly from the peak to zero on day 42. Similarly, the gas production rate in the control bioreactor also decreased quickly from the peak to nearly zero on about day 90. The gas production rate in the 20 g l^{-1} ash added bioreactor appeared to have stayed at a constant rate of about 6500 ml lasting between day 38 and day 51. As a result, the gas production accumulation in the decomposition period of all the bioreactors ranked in the order of 20 g l^{-1} fly ash added bioreactor > control bioreactor > 10 g l^{-1} fly ash added bioreactor > 100 g l^{-1} bottom ash added bioreactor. The lowest gas accumulation happened to the 100 g l^{-1} bottom ash bioreactor was thought to be the higher amount and quicker release of metals hydroxides, carbonates and phosphates. These released OH^{-1}, CO_3^{-2} and PO_4^{-3} etc were considered to give alkalinity buffer and was neutralized by the quick production and large amount of volatile acids generated in the first hydrolysis of MSW anaerobic digestion.

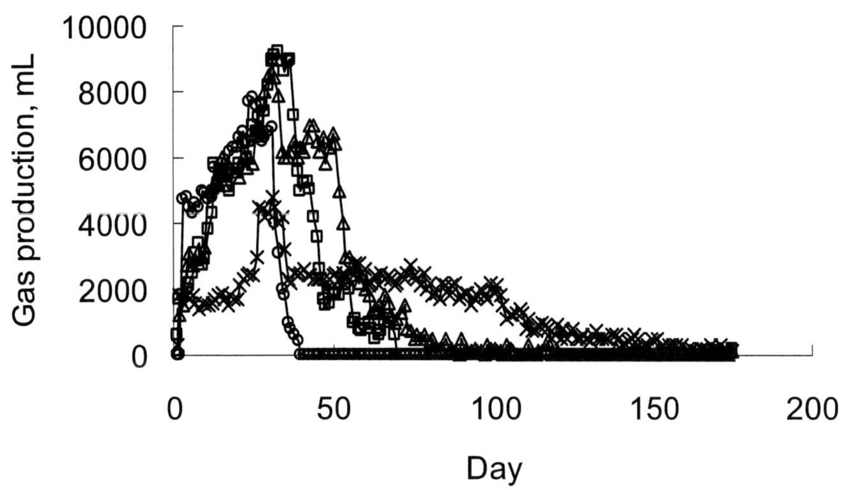

Figure 7. The gas production rate of MSW anaerobic digestion by three different addition ratios of MSWI fly ash in the ash added and the control bioreactors (○: 100 g l-1 bottom ash added bioreactor; □: 10 g l-1 fly ash added bioreactor; △: 20 g l-1 fly ash added bioreactor; □: control bioreactor without fly ash addition).

Evaluation of Anaerobic Bioreactor Performance

From this investigation, the designed ratios of MSWI ashes added bioreactors were found to be able to enhance the gas production rate as compared to the control one. This phenomenon was mainly attributed to the suitable pH, alkalinity and suitable released metals such as alkali metals, trace metals and heavy metals and associated hydroxides and carbonates. Alkali metal compounds and their dissolution of OH^{-1} and CO_3^{-2} were thought to provide the suitable alkalinity and pH offer. In addition, alkali metals were found to be below inhibitory concentration (Parkin and Owen, 1986) and the trace compounds such as dioxin and PAH seemed not exert detrimental effects on bacterial activity and the MSW anaerobic digestion. Similarly, the released heavy metals were also measured to be under the threshold concentration that might be inhibitory on the effects of MSW anaerobic digestion. Further, trace metals were analyzed to be suitable for serving as micronutrients required for the anaerobic microorganisms. These results explained the higher bacterial activity and gas production rate and higher efficiency of MSW anaerobic digestion.

CONCLUSION

The likely utilization of MSWI ashes as landfill cover was examined with anaerobic bioreactor. Four layers were arranged to simulate the co-disposal of MSW and MSWI ashes. From the analysis of leachate parameters and gas production, it was noted that MSWI ashes addition could provide the appropriate alkalinity, adequate released metals and near neutral pH offer. These results particularly with neutral pH 7 and adequate released metal concentration including alkali metals, trace metals and heavy metals associated with the hydroxides and carbonates ensure the suitable and beneficial anaerobic environment for the MSW digestion. The resulting environment suitable for the anaerobic digestion explained the higher gas production rate and stimulatory MSW anaerobic digestion found in the ashes added bioreactor as compared to the control one.

REFERENCES

Ağdağ, O.N., Sponza, D.T., (2005). Effect of alkalinity on the performance of a simulated landfill bioreactor digesting organic solid wastes. *Chemosphere* 59, 871-879.

American Public Health Association, American Water Works Association, Water Environment Federation. (1995). *Standard methods for the examination of water and wastewater.* 19th edition, Hanover, Maryland, US.

Bertolini, L.; Carsana, M.; Cassago, D.; Curzio, A. Q.; Collepardi, M. (2004). MSWI ashes as mineral additions in concrete. *Cement and Concrete Research* 34, 1899-1906.

Bhatti, Z. I.; Furukwa, K. and Fujita, M. (1996). Feasibility of methanolic waste treatment in UASB reactors. *Wat. Res.* 30(11), 2559-2568.

Deschamps, R. J. (1998). Using FBC and stoker ashes as roadway fill: a case study. *J. Geotech. and Geoenviron. Eng.* 124, 1120-1127.

Dichtl, I. N. (1997). Thermophilic and mesophilic (two- stage) anaerobic digestion. *J. CIWEM* 11, 98-104.

Fish, C. *Ph. D. thesis*, Enhance of the anaerobic digestion of municipal solid waste through nutrient supplementation. University of Southampton, Southampton, UK, 1999.

Hjelmar, O. (1996). Disposal strategies for municipal solid waste incineration residues. *J. Hazardous Materials* 47, 345-368.

House, S. J. & Evison, L. M. (1997). Hazards of industrial anaerobic digester effluent discharges to sewer. *J. CIWEM* 11, 282-288.

Kayhanian, M.; Rich, D. (1995). Pilot-scale high solids thermophilic anaerobic digestion of municipal solid waste with an emphasis on nutrient requirements. *Biomass and Bioenergy* 8, 433-444.

Lo, H. M. (2005). Metals behaviors of MSWI bottom ash co-digested anaerobically with MSW. *Resources, Conservation & Recycling* 43, 263-280.

Parkin, G. F.; Owen, W. F. (1986). Fundamentals of anaerobic digestion of wastewater sludges. *J. Environ. Eng.* 112, 867-920.

Paulo, P. L.; Jiang, B.; Cysneiros, D.; Stams, A. J. M.; Lettinga, G. (2004). Effect of cobalt on the anaerobic thermophilic conversion of methanol. *Biotechnology and Bioengineering* 85, 434-441.

Raskin, L.; Stromley, J. M.; Rittmann, B. E.; Stahl, D. A. (1994). Group-specific 16S rRNA hybridization probes to describe natural communities of methanogens. *Appl. Environ. Microb.* 60, 1232-1240.

Takashima, M.; Speece, R. E. (1990). Mineral requirements for methane fermentation. *Crit. Rev. Environ. Control* 19, 465-479.

Zandvoort, M. H.; Gieteling, J.; Lettinga, G.; Lens, P. N. L. (2004). Stimulation of methanol degradation in UASB reactors: in situ versus pre-loading cobalt on anaerobic granular sludge. *Biotechnology and Bioengineering* 87, 897-904.

INDEX

D

E

F

G

H

I

N

O

P

Q

R

S

T

U

V

W

X

Y

Z